AUSTRALIAN SCIENCE
EXPANDING THE FRONTIERS

MANAGING EDITOR **ROBYN WILLIAMS** AM

AUSTRALIAN SCIENCE

CONTENTS

FOREWORD SENATOR THE HON. NICK MINCHIN ... 3
ADVISORY GROUP ... 4
ROLL OF HONOUR ... 6
A MESSAGE FROM SIR GUSTAV NOSSAL ... 7
INTRODUCTION ... 8

PART ONE
AUSTRALIAN ACHIEVEMENTS IN SCIENCE ... 12
Champions of the Past ... 14
The CSIRO ... 28

PART TWO
EXPANDING THE FRONTIERS ... 34
Introduction ... 36
Health and Medical ... 40
Biotechnology ... 74
Marine Science ... 84
Ecology and the Environment ... 90
Agriculture ... 98
Mining and Resources ... 106
Materials Science ... 116
Telecommunications ... 122
Astronomy ... 132
CRCs ... 140

PART THREE
TOMORROW AND BEYOND ... 144
Ethical Issues ... 146
The Next Big Questions ... 148

PART FOUR
VISION STATEMENTS ... 152

PART FIVE
DIRECTORY OF PARTICIPANTS ... 166
CRC Directory ... 184
www.focus.com.au/austscienceleaders ... 186
Index ... 189
About the Writers ... 191

FOREWORD

FOREWORD
SENATOR THE HON NICK MINCHIN, MINISTER FOR INDUSTRY, SCIENCE AND RESOURCES

Australians have always been a resourceful people. Innovation and adaptability are key features of Australia's history.

As Minister for Industry, Science and Resources, I am delighted to see *Australian Science—Expanding the Frontiers* showcase Australian achievements in science, celebrate the high quality of Australia's scientific institutions and research organisations and highlight the vital importance of collaborative investment partnerships between science and business.

The Howard government's $2.9 billion innovation statement—'Backing Australia's Ability'—is the largest funding package for science and innovation in Australia's history. It recognises the enormous contribution science makes to Australia's economic growth, and aims to enable industry and research organisations to continue—even increase—this contribution by taking advantage of the opportunities that increasing globalisation and rapid technological change offer.

'Backing Australia's Ability' encourages research organisations and industry to work more closely together to maintain the world-class standard of our research, and our tradition of international competitiveness, and thus create sustained economic growth for this country.

This publication plays an important role in showcasing the many achievements and exciting advances being made by Australian scientists.

I congratulate all those involved in this exciting publication.

AUSTRALIAN SCIENCE

ADVISORY GROUP

PATRONS

Senator the Hon Nick Minchin
 Minister for Industry, Science & Resources

Sir Gustav Nossal AC, CBE, FAA, FRS
 Professor Emeritus, The University of Melbourne

CHAIRMAN

Sir Laurence Muir
 Director, Focus Publishing Pty Ltd

ADVISORY GROUP

Charles Allen AO
 Chairman, Commonwealth Scientific and Industrial Research Organisation, CSIRO

Professor Warwick Anderson
 Chairman, Research Committee, National Health and Medical Research Council

Professor Peter Andrews
 Co-Director, Institute for Molecular Bioscience, The University of Queensland

Dr Robin Batterham
 Chief Scientist, Commonwealth of Australia

Paul Bell
 President, Human Health Asia Pacific, Merck Sharp & Dohme Asia Pacific

M A (Tim) Besley AO, FTSE
 President, Australian Academy of Technological Sciences and Engineering

Professor Robert Burton
 Director, Anti-Cancer Council of Victoria

Dr Anne Campbell
 Executive Manager, Cooperative Research Centres Association Inc

Professor Adrienne Clarke AO
 Laureate Professor, School of Botany, The University of Melbourne, Ambassador for Biotechnology in Victoria

Professor Suzanne Cory AC
 Director, The Walter and Eliza Hall Institute of Medical Research

Professor David Curtis AC
 Emeritus Professor, John Curtin School of Medical Research, Australian National University

Dr Peter Farrell
 Chairman & Chief Executive Officer, ResMed Inc

Professor Frank Fenner AC, CMG, MBE
 Emeritus Professor, John Curtin School of Medical Research, Australian National University

Professor Peter Gage
 Professor of Physiology, John Curtin School of Medical Research, Australian National University

Dr Annie Ghisalberti
 Director, Questacon, The National Science and Technology Centre

John Grace
 Director, iBIO Pty Ltd

ADVISORY GROUP

Professor Barry Marshall
  Gastroenterologist, QEII Medical Centre, University of Western Australia

Professor John Mattick
  Director, Australian Genome Research Facility, Institute for Molecular Bioscience, The University of Queensland

Dr Nancy Millis AC, MBE
  Emeritus Professor, Department of Microbiology and Immunology, University of Melbourne
  Chancellor, La Trobe University

Dr Jim Peacock AC
  Chief, CSIRO Plant Industry, Commonwealth Scientific and Industrial Research Organisation, CSIRO

Professor Susan M Pond AM
  Deputy Chairman & Director, Johnson & Johnson Research Pty Limited

Steven Rich AM
  Chairman, Focus Publishing Pty Ltd

Steve Skolsky
  Managing Director (July 1999 - April 2001), GlaxoSmithKline Australia

Professor Merilyn Sleigh
  Dean, Faculty of Life Sciences, University of New South Wales

Professor Grant R Sutherland AC, FAA, FRS
  Director, Department of Cytogenetics and Molecular Genetics, Women's and Children's Hospital
  Affiliate Professor, Departments of Paediatrics and Genetics, University of Adelaide

Professor Alan Trounson PhD
  Professor, Obstetrics and Gynaecology, Director Centre for Early Human Development and Deputy Director, Monash Institute of Reproduction and Development, Monash University

Dr Edward Tweddell
  Group Managing Director & Chief Executive Officer, F H Faulding & Co Limited

Professor John White FAA, FRS
  Professor of Physical and Theoretical Chemistry, Australian National University

Robyn Williams AM
  Presenter, ABC Science Show, Australian Broadcasting Corporation

Professor Robert Williamson FRS, FAA
  Director, The Murdoch Childrens Research Institute
  Professor of Medical Genetics, University of Melbourne

Peter Wills AC
  Chairman, CRI Australia Pty Ltd
  Chairman, Health and Medical Research Strategic Review and its Implementation Committee
  Consultative Committee, National Biotechnology Strategy
  Strategic Advisory Committee, Australian Science Capability Review
  Chairman, Garvan Institute of Medical Research (retired May 2001)
  Chairman, Garvan Institute of Medical Research (retired May 2001)

Professor JA Young AO, FAA, FRACP
  Pro Vice-Chancellor (Health Sciences), University of Sydney

AUSTRALIAN SCIENCE

ROLL OF HONOUR

LEAD

AstraZeneca
The University of Sydney

MAJOR

Merck Sharp & Dohme/CSL Ltd

KEY

Allen Arthur Robinson
Anti-Cancer Council of Victoria
CRC for Black Coal Utilisation
CRC for Clean Power From Lignite
GlaxoSmithKline
John Curtin School of Medical Research
Monash University
PricewaterhouseCoopers
Starpharma Pooled Development Ltd
Thomson Marconi Sonar Pty Ltd

CONTRIBUTING

AAMRI
Austin Research Institute
Australian Pharmaceutical Manufacturers Association
Australian Proteome Analysis Facility
Australian Research Council
Australian Technology Park Innovations
Baker Medical Research Institute
Centenary Institute of Cancer Medicine and Cell Biology
Children's Cancer Institute Australia for Medical Research
CRC for Advanced Composite Structures Ltd
CRC for the Biological Control of Pest Animals
CRC for Intelligent Manufacturing Systems and Technologies Ltd
CRC for Molecular Plant Breeding
CRC Reef Research Centre
Eli Lilly Australia Pty Ltd
Garvan Institute of Medical Research
Hanson Centre for Cancer Research
Howard Florey Institute
Macfarlane Burnet Centre for Medical Research
Mater Medical Research Institute
Menzies School of Health Research
Michael Johnson & Associates
Murdoch Childrens Research Institute
Optiscan Pty Ltd
Prince Henry's Institute of Medical Research
Prince of Wales Medical Research Institute
Redfern Photonics/Australian Photonics CRC
St. Vincent's Institute of Medical Research
State Government of Victoria— Science, Technology & Innovation
The Mental Health Research Institute of Victoria
The Queensland Institute of Medical Research
The University of Melbourne
The Walter and Eliza Hall Institute of Medical Research
TVW Telethon Institute for Child Health Research

INTRODUCTION

FOREWORD

PROFESSOR EMERITUS SIR GUSTAV NOSSAL AC CBE FAA FRS

It is with great pleasure that I present *Australian Science—Expanding the Frontiers*, a book that celebrates scientific excellence in Australia. This country has a long tradition of scientific achievement—the quality of Australia's scientific institutions and research centres, plus the extraordinary intellectual capital of its scientists and researchers, are the key factors in this. The book describes the integral role our scientists have played—and continue to play—in the growth and development of the Australian economy, and in worldwide scientific endeavour.

One of the major tasks facing Australia today is to foster and nurture our scientific intellectual capital. This book will help the scientific community improve awareness in business and the wider community of the talent and resources that exist here.

The book also shows that our world-class facilities, infrastructure and resources make Australian research and development-based industries outstanding investment opportunities. Investment partnerships between science and business are vital, as they create the essential connection that allows research to progress to the international market. This book provides recognition for many of the companies and organisations already doing this.

It shows that Australian science is like Australia as a country: big, exciting, innovative and brave, but also thoughtful, practical, business-minded and dogged. From this wonderful mixture have come many breakthroughs, and many still lie ahead. I commend the book to all Australians, both within and without the science and business communities.

AUSTRALIAN SCIENCE

INTRODUCTION

SCIENCE IN AUSTRALIA HAS A LONG HISTORY—60,000 YEARS!—AND, IT SEEMS, A BRIGHT FUTURE, JUDGING BY THE PEOPLE'S ENTHUSIASM AND THE POLITICIANS' COMMITMENT, REPORTS ROBYN WILLIAMS

At the beginning of 2001, as if to coincide with the centenary of Federation, science in Australia took off as an issue. The prime minister launched the Innovations Statement 'Backing Australia's Abilities' in what was heralded as the largest single budget for research and development (R&D) in the nation's history. At the same time the Leader of the Opposition proclaimed 'The Knowledge Nation', saying that brains are the key to Australia's future.

To cap it all, a national survey, published on Australia Day, showed that people put science and technology above politics, even sport, as the most important key to our prosperity in the 21st century.

What does this mean in terms of policy? Should we try to be good at everything—surely a stretch even for a very large country such as ours? Or perhaps risk picking winners—a research practice generally thought to be as perilous as racing fleas.

Essentially, it means we will continue our work in the fields which underpin our major interests, then add some of the activities from the high frontier of science. Both state and federal governments will, henceforth, endeavour to foster scientific enterprise as never before.

Science policy in Australia, odd as it may seem, is barely as old as Kylie Minogue. During most of the 20th century it was assumed that R&D happened 'overseas' and we could simply add our own bits. When, during World War II, we were cut off from international sources of instrumentation, it took Australia barely a few months to start producing a staggering array of original devices. After the

INTRODUCTION

war, as foreign connections were re-established, this stream of inventiveness was blithely shut down.

Of course we did have various scientific institutions, and their funding was the extent of our science 'policy'. This began to change in the 1970s, with ministers Malcolm Fraser and Bill Morrison. A Science Council was, at last, established. Then, in the 1980s, following the publication of his book *Sleepers Wake*, science minister Barry Jones trumpeted the connections between ideas, innovation, real wealth and social progress. It was at this time that Professor Ralph Slatyer was appointed as our first Chief Scientist. It was he who managed to persuade his old school friend, prime minister Bob Hawke, of the merits of Cooperative Research Centres (CRCs).

These CRCs (listed in this book) combine the intellectual and enterprise talents of universities, institutions such as CSIRO, and business. They are funded by government, but for a limited period. A mark of their success, as Dr Bruce Cornell (see the biotechnology chapter) says, is to be able to say: 'We're off and running—we don't need that support any more.' Their establishment is, in part, an acknowledgment of the time that often elapses between idea and product, between need and result, between 'what if?' and 'a-ha!' in the world of science. The degree and amount of both hope and faith required to pursue science research is enormous, perhaps greater than in many other fields, because the possibility of failure is extraordinarily high. However, the successes are equally extraordinary. It is this nexus between long-term investment, by both government and the private sector, and risk and benefits and profits, that the CRCs are a response to.

Barry Jones also invented The Australia Prize. This was to echo the Japan Prize or even the Nobel. Offered for a different subject each year (starting with GM plants!) the prize has transmuted, after 10 years, into The Prime Minister's Science Prize (given in 2000 to Jim Peacock and Liz Dennis for GM: the genes for flowering).

Having the prime minister head the prize was an initiative recommended at the December 1999 meeting of the PM's Science Council. We (I took part) said that science urgently required leadership at the highest level. Only this would signal its significance to all portfolios in government. Only this would show the public the extent to which science underpins every activity, from farming to sport, from leisure to law.

There are other prizes, such as the Eurekas, centred on the Australian Museum in Sydney, and

AUSTRALIAN SCIENCE

the Clunies Ross Awards, based around the Academy of Technological Sciences in Melbourne; both have boosted recognition of creativity and efforts to promote scientific ideas.

Both scientific creativity and the promotion of science ultimately depend on resources. The major source of grants nationally is the Australian Research Council (ARC). Until recently the Council told all comers that only one in five of the excellent proposals it received could be funded. Now the federal government has doubled the contents of ARC's treasury. Schemes have also been set up to encourage our brightest research people to return to these shores. One who would like to be tempted to return is Dr Brian Gaensler, the astrophysicist at MIT (Boston) voted Young Australian of the Year in 2000. Bright minds benefit from experience abroad, but it is essential to give them good reason to come back.

Each state also has its own scheme to promote enterprise. The Australia Technology Showcase of New South Wales, with 300 company members, will soon go national. Queensland is backing its already successful biotechnology and marine science research and industry. Victoria has a proud record across the board, especially in small business. South Australia favours the wine industry (spectacularly), as does Western Australia—and both conduct major research in minerals.

When we look at the size and variety—in landscapes and plant and animal life—of the Australian continent, it should be no surprise that traditionally, Australian science that is linked to agriculture, mining and the ocean has always been of the highest quality. Australia has arid areas and alpine areas, wetlands, a vast coastal fringe, islands, atolls and even a large piece of Antarctica to manage. It has a huge tropical area (38 per cent of the landmass), and large tracts of land that only became arable when Profesor Eric Underwood identified zinc deficiency as the problem. And now, 25 per cent of our arable land has been seriously, perhaps permanently, degraded by inappropriate use over the last 200 years. It is easy to see why agtriculture research, now augmented by the latest advances in genetic engineering, remote sensing and conservation, has remained a priority.

Space research, which has been important from the very start of the colony with the efforts of Sir Thomas Brisbane and then John Tebbutt, is another continuing fundamental. It is no coincidence that the renowned Robert Hanbury Brown, professor of astronomy at the university of Sydney, one of the 'radar boys' and pioneer of stellar interferometry, is actually the original 'Boffin'.

INTRODUCTION

As for natural history, it is extraordinary to recall that two expeditions from Britain were launched to Australia even in the 20th century: one was to the Great Barrier Reef (in the 1920s) and one was to the Kimberley (in the 1980s). Much remains undiscovered. Knowing what is there is the key to conservation.

Most of the fresh fields actually stem from our traditional areas of expertise. Medicine, for instance, has always been important, not least an understanding of tropical diseases. Some people are surprised to learn how much of Australia is in the tropics. Will the amount of our country that is tropical increase as global warming bites?

Telecommunications has also always been important in Australia to overcome the 'tyranny of distance'. Marconi sought out this nation for one of his radio links early in the 20th century. Now the federal government has named ICT (information and communications technology) as one of our favoured specialties. The challenge is to link ourselves effectively, especially in broadband, with the huge nations in our region: the 'Challenge of Our Time Zone!'

Energy and materials science also emerge from our roots. Is it surprising that Martin Green should lead a world-beating solar technology team when you think of the full force of sunlight blasting down on this continent every day of the year?

And as for the range of minerals and other ingredients that are found in this country, and from which we can conjure almost any material—just imagine: on my desk in front of me is a stone, ordinary enough to look at, which was lifted from somewhere in Western Australia. It is a piece of zirconite, officially dated at 4.2 billion years old. That's nearly as old as the solar system itself!

With such a basis it is hard to see any real limit to the high frontier of science in Australia. But it does depend: it is a matter of culture.

Science is not an 'add on'. It is not the field you opt for instead of real estate or videogame.com. Australia, in 2001, as our political and business leaders may now realise, is a place where we need some sense of future, some vision of what we want our nation and our region—indeed our world—to be like in the 21st century. Science and technology are the vehicles that will help us achieve such a vision. Both are essentially international—the first and most fundamental 'globalised' activities.

We have the ingredients, it is now time to get on with it.

PART ONE
AUSTRALIAN ACHIEVEMENTS IN SCIENCE

A university chemistry class in 1896
> RIGHT: The chemistry laboratory at Sydney University, used from the 1880s

CHAMPIONS OF THE PAST

CHAMPIONS OF THE PAST

AUSTRALIA'S EUROPEAN HISTORY HAS ALWAYS INVOLVED SCIENCE—OUR STRANGENESS AND OUR ISOLATION HAVE BOTH MADE SCIENTIFIC STUDY NECESSARY, REPORTS BERNADETTE HINCE

Any short account of Australian contributions to science over the past two centuries must choose only parts of what is a huge story to tell. I have picked out some strands from this very large arena. This chapter looks only at the European Australian (or 'neo-European') scientific world, and does not cover relevant indigenous knowledge and views.

In August 1768, the ship *Endeavour* sailed from England under the command of James Cook. English astronomers had calculated that in Tahiti on 3 June 1769 the transit of Venus across the face of the sun would be visible, and the Royal Society of London had conceived the expedition to help solve the problem of determining longitude. But the voyage had another powerful patron and another objective: on secret instructions from the British Admiralty, Cook was to sail west from Tahiti to explore the South Pacific for a Great South Land. In following these instructions, Cook made his way southwest from Tahiti, eventually sighting Australia's east coast in April 1770. Cook was a skilled navigator and observer, but he also had with him on the *Endeavour* several natural historians and artists. Joseph Banks, Daniel Carl Solander, Hermann Spöring and Sydney Parkinson botanised and collected specimens at every possible opportunity. The link between two important endeavours, astronomical and colonial, was forged at Possession

AUSTRALIAN SCIENCE

PICTURE: UNIVERSITY OF SYDNEY ARCHIVES

> **ABOVE:** A laboratory in Sydney University's Medical College

other explorers. Members of various French expeditions—*La Recherche*, *La Naturaliste*, *L'Uranie* and *La Boussole* and *L'Astrolabe*—made notable collections and illustrations (Finney, CM [1984] *To sail beyond the sunset: natural history in Australia 1699–1829*: 2). Between 1801 and 1803 Matthew Flinders (with botanist Robert Brown and artist Ferdinand Bauer) circumnavigated the continent. When Flinders' ship, the *Investigator*, first landed at King George's Sound in Western Australia, the men collected about 500 new species of plants (Moyal A [1986] '*A bright and savage land*': *scientists in colonial Australia*: 21).

The country's flora and fauna were radically different from those of the northern hemisphere,

THE FLORA AND FAUNA OF AUSTRALIA WERE RADICALLY DIFFERENT FROM THOSE OF THE NORTHERN HEMISPHERE, AND STRUCK THE EARLY EUROPEAN SETTLERS AS BIZARRE AND UNWELCOMING

Island, off the coast of Queensland, when Cook claimed the continent for Britain.

Cook was not the first European to visit these shores, but encounters with the Australian coast by previous Dutch, Spanish and Portuguese explorers have left few records. In 1699 William Dampier, visiting in the *Roebuck*, made lengthy reference to the plants and animals he encountered on the arid western coast of the continent. But in his and others' travels and enquiries, there was little communication with the country's indigenous occupants: there was no common language.

The plant, animal and mineral specimens that Cook and his companions took back to England on the *Endeavour* were added to by those of

and struck the rest of the world, and the early European settlers, as bizarre and unwelcoming. Accounts from the early 19th century talk of the country's wooden pears, scentless flowers and unmusical birds (Drayson N [1997] 'A land of contrarieties: 19th century natural history writing in Australia'). Nevertheless, the new country's natural resources and environment were generally regarded unquestioningly as a rich lode for scientific enquiry. When the flora, fauna and mineral products of Australia began to reach Britain and Europe, they caused sensations in the scientific world. Some elements of Australia's mammalian fauna—foremost among them the platypus—were outrageous enough to European eyes to lead to speculation that specimens were cunning taxonomic hoaxes. Well before structures to facilitate scientific growth and discovery had been created in Australia, its 'natural productions' were already making their first impact on the world—the first impact of Australian science was an impact on world science.

By the early 19th century, natural history was at the height of its popular appeal in Europe. Australia was a bonanza for naturalists, and in the late 1820s her colonial populations began creating their own institutions for scientific enquiry. Increasingly, as the century drew on, the newly established museums of natural history, geology and mining, and applied science, as well as agricultural and learned societies, observatories, herbariums and universities, began the task of cataloguing the natural history of Australia in Australia. At the same time, inland journeys of exploration, and geological and other scientific surveys, were bringing new collections to the repositories of these institutions. But though such journeys of exploration had specific purposes, often economic, coverage of the vast continent was essentially random as far as science was concerned.

>> SCIENCE INSTITUTIONS AND TEACHING

The mid to late 19th century saw the development of a formal world of science in Australia. Geologist and mineralogist Archibald Liversidge (1846–1927) energetically promoted the cause of science in the country. Liversidge came to Australia from London to a geology appointment at the University of Sydney. He helped establish the Australasian Association for the Advancement of Science (later the Australian and New Zealand Association for the Advancement of Science) in 1888, and the splendidly named Industrial, Technological and Sanitary Museum in Sydney.

Liversidge was founding dean of Sydney University's Faculty of Science in 1879, for which he fought 'Homeric battles with the forces of Arts' (*Australian Dictionary of Biography* [1974], vol 5: 93), though science had been taught in Australian universities since at least 1855 by people such as Frederick McCoy (1817–99), first professor of natural history at the University of Melbourne. Charles Darwin published *The origin of species* in 1859. From the 1860s through to the 1880s McCoy and others in influential teaching positions within the colonies opposed the Darwinian argument, but succeeding scientists espoused it.

Anthropologist, biologist and supporter of the Australian arts Walter Baldwin Spencer (1860–1929) became foundation professor of biology at the University of Melbourne in 1887. Spencer—whose teachers at Manchester and Oxford had strong Darwinian leanings—contributed energetically to the expansion of science through his brilliant teaching, and through his support for the Australasian Association for the Advancement of Science, and Victoria's Field Naturalists Club and Royal Society (*Australian Dictionary of Biography* [1990], vol 12: 33–36).

Scientific societies played an important part in establishing the role of science in Australia. There had been early attempts to create philosophical societies in both New South Wales

AUSTRALIAN SCIENCE

and Tasmania. The creation of Royal Societies was more successful. The Royal Society of Victoria was founded in 1854, with Victorian government botanist Ferdinand Jakob Heinrich von Mueller (1825–96) as its first president. In 1857 von Mueller became director of Melbourne's Royal Botanic Gardens and, in the tradition of the times, exchanged Australian seed and plants widely throughout Europe and North America. Institutions and associations such as the Royal Societies and Geological Surveys embodied the growing independence of Australian society. As historian Ann Moyal noted:

> By the [19th] century's end, Australia still depended substantially on British technology, concepts and instrumentation for her research needs, but she was shaping a maturing scientific community of her own. (1986: 13)

In the early 1950s, Marcus Oliphant and David Martyn lobbied the government to establish a national scientific society. They were successful, and the Australian Academy of Science was founded in 1954, with Oliphant serving as its first president. At the core of the young academy were Martyn and other Australian Fellows of the Royal Society. Martyn's skill in negotiating was fundamental to the realisation of the Academy, and Oliphant was a brilliantly successful fundraiser. After the Academy was founded, his approaches resulted in a donation of £25,000 from Broken Hill Proprietary Ltd for a building to house the Academy in Canberra.

Today the Academy is made up of Australia's elite scientists, from all fields, and membership is by election. It is now an independent body, though it receives government grants. The Academy has always had close links with industry, and the nexus between science and industry is one of its five main areas of concern. It was partly as a result of Oliphant's later efforts that in 1972 the Australian government established an advisory committee on science and technology.

David Martyn (1906–70), Oliphant's collaborator in pushing for an Australian Academy of Science, was a Scottish-born physicist who came to Australia to take up a position with the Council for Scientific and Industrial Research's Radio Research Board.

Martyn proposed an upper atmosphere which differed radically from the accepted view. He investigated solar radiation, work which fed into the refinement of radar during World War II, and coined the phrase 'quiet sun' to describe times of low sunspot activity. Australia was consequently able, during World War II, to develop considerable expertise in radar, which not only enhanced Australia's contribution to the war effort, but affected the directions of research after the war (Bolton HC [1990] 'Optical instruments in Australia in the 1939–45 war: successes and lost opportunities', *Australian Physicist* 27(3)).

Martyn was founding chief (1941–42) of CSIR's Division of Radiophysics. As a result of his world eminence in ionosphere and radio astronomy research, he became president of the Radio Astronomy Commission of the International Union of Radio Science in 1950, then of the Union's Ionospheric Commission, and chairman of a subcommittee of the UN Committee on the Peaceful Uses of Outer Space.

CHAMPIONS OF THE PAST

PICTURE: UNIVERSITY OF SYDNEY ARCHIVES **PICTURE:** AUSTRALIAN ACADEMY OF SCIENCE

> **ABOVE LEFT TO RIGHT:** Prof. Archibald Liversidge (1847–1927); David Martyn (1906–70), founding chief of CSIR's Radiophysics Division

>> GOING SOUTH—AUSTRALIAN ANTARCTIC EXPLORATION AND RESEARCH

A preoccupying aim in 19th and early 20th century Antarctic exploration and science was reaching the south geographic and south magnetic poles. In 1831 the Englishman James Ross had reached 89°59'N, effectively the north magnetic pole, and in 1909 Robert Peary, Matthew Henson and four companions had reached the north geographic pole. Between these two dates, knowledge of electromagnetism of the Earth had advanced hugely, sunspot activity had been related to magnetic storms, Guglielmo Marconi had sent radio signals across the Atlantic, and Oliver Heaviside had theorised the existence of an atmospheric layer (later named the ionosphere) which bounced radio waves back to Earth, enabling long-distance transmission across its curved face.

In the 19th century, Australians, including von Mueller (a member of the first Australian Antarctic Exploration Committee), were interested in the Antarctic continent for scientific, exploration and economic reasons. Australian explorers were at the forefront of early scientific discoveries in Antarctica. Among them was explorer and geologist Douglas Mawson (1882–1958). Mawson had come to Sydney as a young child with his parents from Yorkshire in England; he studied engineering and geology at the University of Sydney and was the physicist on Ernest Shackleton's 1907–09 Antarctic expedition. On 16 January 1909, Mawson, TW Edgeworth David and physicist and biologist Alistair Mackay were the first to reach the region of the south magnetic pole. Mawson was an important Australian contributor to the expression of scientific nationalism in the Antarctic regions. He made a second voyage south, leading the first Australian expedition to the Antarctic in 1911–14. On this trip he survived one of the most extraordinary lone journeys in inland polar travel after both his

AUSTRALIAN SCIENCE

PICTURES: UNIVERSITY OF SYDNEY ARCHIVES

> **ABOVE LEFT TO RIGHT:** Sir Douglas Mawson (1882–1958); Prof. TW Edgeworth David (1858–1935) reports on Antarctica 1909

companions died. In later surveying, Mawson identified the radioactive element radium at Mt Painter in the Flinders Ranges of South Australian .

>> AGRICULTURE

Much of the impetus for agricultural experimentation and development in Australia, in both the 19th and 20th centuries, came from the need to find crops and animals which were appropriate or could be adapted to the country's generally harsh conditions. This led to innovative machinery, and to selection and cross-breeding experiments with wheat which were the first systematic plant breeding experiments to apply the new genetic theories of Gregor Mendel.

Plant breeder William Farrer (1845–1906) came to Australia from England at the age of 25, in poor health. He worked in New South Wales as a surveyor, and began a series of breeding experiments with wheat, creating many varieties that were particularly well suited to Australian conditions. Farrer's most well-known wheat variety, 'Federation', was widely grown in the early part of the 20th century, during which time Australia changed from a net importer of wheat to a net exporter. The extraordinary success of Federation was mainly because it was high yielding and so drought resistant it enabled considerable expansion of the wheat belt inland. Because it was early maturing it was less vulnerable to rust, a fungal disease which is more of a problem in wet conditions.

>> GEOLOGICAL JIGSAWS AND MINERAL WEALTH

Australia's geology has always been of great interest to both Australian and Northern Hemisphere

CHAMPIONS OF THE PAST

scientists, who used Australian data to test some European theories. One such theory was that of 'continental drift', which sought to explain the way coastlines of countries and continents apparently dovetailed with each other. An Australian geologist, the Reverend William Branwhite Clarke, saw geological similarities between the coastline of Western Australia and the east coast of Africa.

An explanation for this similarity, the theory of continental drift described by Alfred Wegener in 1912, has since been confirmed by modern geological, palaeomagnetic and seafloor studies, and much of it survives today in modified form as the modern theory of plate techtonics.

The discovery of gold in New South Wales and Victoria in the 1850s was a huge impetus for the formalisation of science in Australia. It led to the establishment of colonial geological surveys, as well as providing a massive source of government revenue and hence capital works. Between 1852 and 1899, geological surveys were established in each colony to explore and map geological resources.

After Federation in 1901, both federal and state governments offered large rewards or subsidies for the discovery of new mineral resources: £50,000 in 1920 for the discovery of commercial oil, up to £25,000 tax-free in 1948 for economically viable deposits of uranium (Wilkinson R [1996] *Rocks to riches: the story of Australia's national geological survey*: 386–88). During World War II the national Mineral Resources Survey was formed to assess Australia's reserves of important minerals and petroleum deposits. Immediately after the war, in 1946, its successor, the Bureau of Mineral Resources, was constituted. Alongside state surveys, it became active in systematic geological field mapping surveys from the late 1940s to the 1970s. Though there had been proposals for the creation of a Commonwealth geological survey since soon after Federation, it had taken the impetus of war to establish one.

PICTURE: UNIVERSITY OF SYDNEY ARCHIVES
> **ABOVE:** Geology expedition—Farey Camp in 1880s. Professor TW Edgeworth David (left).

>> MEDICAL RESEARCH

One of Australia's science strengths in the 20th century has been in the field of medical research, particularly immunology. During the early and mid-20th century, in medicine as in other scientific fields, there was little local specialist training available. Australian scientists went abroad to train—largely to England—as a matter of routine.

The dominance of Europe and England was reflected not only in movements away from

Australia, but also in movements towards it. The first director of the Walter and Eliza Hall Institute of Research in Pathology and Medicine, Sydney Patterson, was English. He was followed by medical scientist Charles Kellaway who, although Melbourne-born, had been working in London at the National Institute of Medical Research after his discharge from the Australian Army. The Walter and Eliza Hall Institute was founded in Melbourne in 1915, using money from the trust set up by Eliza Hall to perpetuate the memory of her late husband, businessman Walter Hall. Much of the Institute's initial work concerned toxins and venoms—Kellaway worked on the pharmacology of snakes, spiders and mussel venoms—and parasites such as hydatids. It later established a leading reputation for virology and immunology.

rickettsia virus which was named *Coxiella burnetii* in his honour. He concentrated on the influenza virus in his later research. Burnet returned to England in 1932, where he studied the immunology of viral infectivity and the inhibition of viruses by antibodies at the National Institute for Medical Research. In 1960, together with Peter Medawar of the United Kingdom, Burnet won the Nobel Prize for Medicine for work on immunological tolerance. Burnet is one of five Australians to win Nobel Prizes for Medicine, 'an unreserved patriot who refused overseas appointments which could have furthered his career' (Golden Press [1986] *The Concise Australian Reference Book*: 334).

Burnet was director of the Institute from 1944 until 1965. During this time, many significant researchers were appointed to the Institute staff,

ONE OF AUSTRALIA'S SCIENCE STRENGTHS IN THE 20TH CENTURY HAS BEEN IN THE FIELD OF MEDICAL RESEARCH, PARTICULARLY IMMUNOLOGY

One notable researcher at the Institute was immunologist and virologist Frank Macfarlane Burnet (1899–1985), who began work there as a registrar in pathology in 1923. Kellaway put him in charge of the bacteriology department, then sent him to London, where he worked at the Lister Institute on bacteriophages, work which he continued at the Walter and Eliza Hall Institute from 1927. He studied the deaths of 12 children after a 1928 Queensland inoculation program, worked on Q (Queensland) fever, psittacosis, poliomyelitis, herpes and the ectromelia virus, and demonstrated that Q fever was caused by a

including gastroenterologist Ian Wood, microbiologist Frank Fenner, Gordon Ada, and US researcher (Daniel) Carleton Gajdusek, whose work at the Institute in the 1950s and early 1960s on the origin and dissemination of the infectious degenerative brain disease kuru won him the 1976 Nobel Prize for Medicine. Kuru—which occurs only in the Fore people of the eastern highlands of Papua New Guinea—is related to scrapie, bovine spongiform encephalopathy (BSE, also known as 'mad cow disease') and Creutzfeldt-Jakob disease. Another immunologist, Gustav Nossal (1931–), succeeded Burnet in 1965 as director of the Institute, expanding the research to include experimental haematology, diabetes, transplant biology and the origins and behaviour of cancer. He has since been succeeded by molecular biologist Suzanne Cory (1942–).

Howard Florey (1898–1968) was the medical scientist responsible for discovering one of the most fundamental medical findings of the 20th century: the antibacterial properties of penicillin.

Florey went to Oxford in 1922 as a Rhodes scholar. He worked in Cambridge and London; there, in 1928, Alexander Fleming observed the growth of a fungus with antibacterial properties in a petri dish left on a bench. That fungus turned out to be *Penicillium*.

Florey's studies on mucous secretion in the gut led him to wonder why the wall of the gut was relatively resistant to infection by bacteria, and this aroused his interest in lysozyme, a bacteriolytic enzyme that had been discovered by Alexander Fleming in 1920. In order to investigate this problem further, Florey obtained a grant to employ a chemist to purify lysozyme, and persuaded E.B. Chain to study its substrate. This led Florey and Chain to discuss a broader project, namely the study of the many antimicrobial substances known to be produced by bacteria and fungi. He chose penicillin as the first agent to investigate (Fenner [1998] *Portrait of Howard Florey*).

World War II provided an impetus for important contributions to medical research. In 1939 Florey and Chain headed a team of British scientists who, with funding from the Rockefeller Foundation, extracted penicillin in small amounts from a liquid

PICTURE: AUSTRALIAN ACADEMY OF SCIENCE

> **ABOVE:** Sir Macfarlane Burnet (1899–1985)

culture. With government backing, penicillin was made available to Allied troops by the time of the Normandy landings (Fenner 1998).

Florey was the first Australian to be elected president of the Royal Society, and the first Australian to receive the Nobel Prize for Medicine, awarded in 1945 'for the discovery of penicillin and its curative effect in various infectious diseases'. He was instrumental in establishing the John Curtin School of Medical Research at the Australian National University (ANU) in 1953 and served as its

AUSTRALIAN SCIENCE

> **ABOVE:** Sir Howard Florey (1898–1968)
PICTURE: JOHN CURTIN SCHOOL OF MEDICAL RESEARCH

first head. In 1947 the Howard Florey Institute of Experimental Physiology and Medicine was set up at the University of Melbourne, with funding from the US Rockefeller Institute and the National Institutes of Health, as well as from Australian donors.

Medical research had by then long been a forte of Australian science. For example, it was parasitologist and naturalist Thomas Bancroft (1860–1933) who discovered that the mosquito *Aedes aegypti* carried dengue fever:

> *His generosity in providing free access to materials for colleagues enabled identification of many new species from such diverse groups as freshwater algae, eucalypts, mosquitoes, spiders, snakes, fruit-flies and fish. Today, Bancroft's name lives on in many of these species: a fitting tribute to this generous and ... talented man.* (Australian Institute of Political Science [2000] *Tall Poppies 2000*)

The crossover between animal and human research was further exemplified by the development of in vitro fertilisation (IVF): 'In 1960 Neil Moore of the Department of Animal Husbandry, University of Sydney, was the first [in the world] to carry out embryo transfer (in sheep) and in 1976 froze cattle embryos for export' (Golden Press 1986: 340). The techniques now used in human IVF are based on Moore's pioneering work. Among very early results in relation to human research was the birth in Melbourne in 1980 of Candice Reed, the fourth 'test-tube baby' in the world, born two years after the first successful births in Britain.

Physician and immunologist (Annie) Jean Macnamara (1899–1968) was the first woman resident medical officer of the Melbourne Children's Hospital, in 1923. During the 1920s and 1930s there were epidemics of poliomyelitis in Australia. Macnamara researched polio using serum from polio victims, testing immune serum therapy on her own patients at the pre-paralytic stage. Working with Macfarlane Burnet, she discovered that there was more than one type of polio virus. Her findings were critical to the later development of the Salk vaccine. In 1931 she went to England and the United States on a Rockefeller Fellowship to study the polio virus. Macnamara worked in private practice with polio patients, and at the Walter and Eliza Hall Institute in Melbourne.

Macnamara's near-contemporary Elizabeth Kenny (1880–1952) also made a significant medical contribution with her unorthodox treatments for polio patients. The methods were controversial, and a 1938 Queensland Royal Commission rejected them. In 1940 she moved to the United States, where her methods were accepted widely. Her treatment was successful in rehabilitating many.

During her time in the United States, Macnamara had met Richard Shope at Princeton University. Shope was working on fighting the myxoma virus in rabbits. Macnamara encouraged its introduction into Australia to control rabbit populations by corresponding with the Australian High Commissioner in London (Stanley Bruce), which led to CSIR trials of the virus in the 1930s and 1940s. Though these trials were not sufficiently successful, Macnamara called for further trials. Those on the Murray River established a mosquito as vector (carrier) for the virus, and led to outbreaks of myxomatosis in 1950–51 which dramatically reduced rabbit populations.

John Eccles (1903–97) graduated in medicine from the University of Melbourne in 1925, winning a Rhodes scholarship to Oxford University, where he studied the transmission of impulses in the nervous system. Eccles shared the 1963 Nobel Prize for Medicine with Alan Hodgkin and Andrew Huxley (Britain) for work the three did at Oxford in the 1930s. They discovered the mechanisms responsible for chemical transmission of nervous impulses. Eccles returned to Australasia, and worked in New Zealand on the chemical and electrical properties of synapses, before becoming Professor of Physiology at the ANU in Canberra.

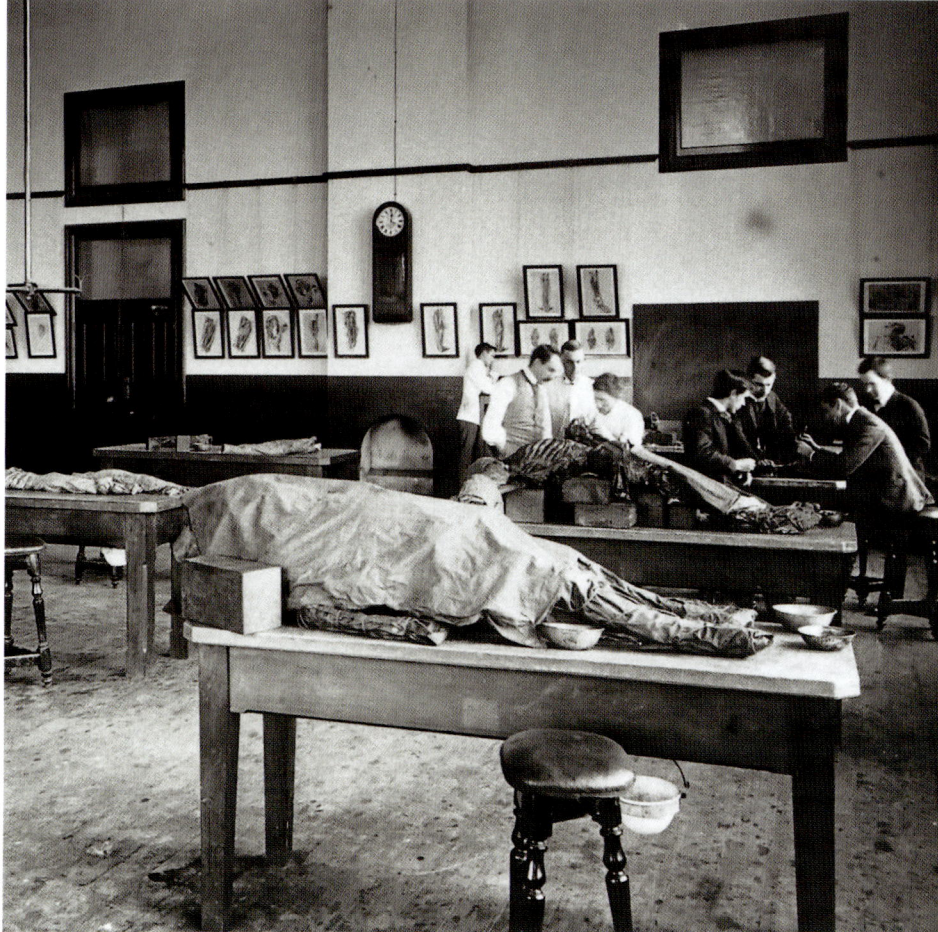

PICTURE: UNIVERSITY OF SYDNEY ARCHIVES
> **ABOVE:** Dissecting room at the University of Sydney, 1900–10

A public figure of great style was ophthalmologist Fred Hollows (1929–1993), who was born in Dunedin, New Zealand, and came to Australia in 1960. Hollows helped establish the first Aboriginal medical centre, treated trachoma in indigenous Australians, and inspired others to join him. In 1976 he set up the National Trachoma and Eye Health Program, which operated clinics in remote communities in Australia.

Using money donated by the Australian public, Hollows established clinics and factories to manufacture lenses in places as widespread as Eritrea, Nepal and Vietnam, and trained local doctors to treat avoidable blindness, including performing basic surgery for cataracts.

AUSTRALIAN SCIENCE

> **ABOVE:** Sir John Eccles (1903–97)
PICTURE: AUSTRALIAN ACADEMY OF SCIENCE

>> MINING AND NUCLEAR TESTING GROUNDS

In February 1932, the Englishman James Chadwick announced the discovery of the neutron. Twenty-one years later, the Australian Atomic Energy Commission, with a small research facility at Lucas Heights, south of Sydney, was established to promote exploration for uranium and nuclear energy research (Wilkinson 1996: 390–91). Uranium deposits had already been found in South Australia and the Northern Territory. Increasing interest in uranium, fostered by government rewards, led to the discovery of rich deposits at Rum Jungle in the Top End of the Northern Territory in 1949. The claim was immediately taken over by the young Bureau of Mineral Resources, which assessed its economic potential, and it was this mine which eventually supplied uranium to Britain's atomic test program until 1971. Another mine, Mary Kathleen in northwest Queensland, produced uranium from 1958 to 1963 and again from 1976 to 1982.

The scientific legacy of World War II was substantial. The war (in particular, the effects of the refinement of radar) strengthened links between defence and civilian science, as well as reinforcing research links between Australia and Britain. Nuclear weapon and guided missile research after the war led to a request from Britain for land on which to conduct tests of these weapons. In 1947 the Australian government made land available at Woomera, in northern South Australia:

The first test was held in October 1952 in the Monte Bello Islands off the coast of Western Australia. The following year two more atomic devices were exploded at Emu Field, part of the Woomera Rocket Range in South Australia. The first of these, Totem I, is thought to have been responsible for the 'black mist'—a mysterious cloud that descended upon Aboriginal communities to the northeast of the test site, causing vomiting, diarrhoea, skin rashes and sore eyes. The long-term health effects have never been determined (Sherratt T [1996] 'On the beach: Australia's nuclear history', paper presented at the French Worlds, Pacific Worlds Conference).

Tests at Maralinga, another weapons testing site in South Australia, dispersed some 22 kg of highly radioactive plutonium over about a million square kilometres of 'virtually uninhabited' Australian desert country. Plutonium testing lasted until 1963, and the Woomera range tested British weaponry until 1980 (Golden Press 1986: 336, 346). The Australian Defence Scientific Service, later the Australian Nuclear Science and Technology Organisation, became Australia's second-largest scientific research organisation.

The atomic age had another side. After the bombing of Hiroshima and Nagasaki in Japan at the end of World War II, Bertrand Russell and Albert Einstein called on scientists around the world to meet and discuss their concerns about the use of thermonuclear weapons. The Australian representative at the first meeting, a gathering of 22 scientists from around the world in Pugwash, Nova Scotia, Canada, was Mark Oliphant.

Oliphant (1901–2000) had taken up a scholarship at the Cavendish Laboratory in Cambridge, under its New Zealand-born director, the nuclear physicist Ernest Rutherford. There Oliphant combined practical ability with his research interests, building an accelerator to study the effects of high-speed protons on lighter elements, used the newly discovered isotope deuterium, and discovered a light form of helium (helium-3). His findings underlay the understanding of both fusion and fission physics, enabling the use of thermonuclear energy in weapons during World War II.

From Cambridge, Oliphant went to Birmingham in 1937 to found a nuclear physics laboratory. The resonant cavity magnetron, invented in the early 1940s in Oliphant's laboratories by John Randall and Henry Boot, allowed radar to operate at shorter wavelengths than previously possible, and was deployed in the war.

Oliphant spent the years 1943–46 in Berkeley, involved with the Manhattan project, whose aim was to develop fission nuclear weapons, and working on a method for separating the uranium isotope U235 from isotope U238. He then returned to Australia and was founding director of the Research School of Physical Sciences at the new Australian National University in Canberra. He became Governor of South Australia in 1971, and remained in the position until 1976.

>> REACH FOR THE STARS

European occupation of Australia coincided not only with a great flowering of interest in natural history, but with a time when an intense interest in the heavens was accelerated by rapid advances in optical technology. Australia offered a view of a different sky. It also offered a place for observing this sky in clear conditions, and Australian astronomers soon began to find a role in the international world of astronomy.

Australia-born John Tebbutt (1834–1916) pursued his interest in astronomy from an early age. By the age of 27 he had discovered a comet (later 'Tebbutt's comet') and two years later he had built his own observatory. He was the first person

CHAMPIONS OF THE PAST

WORLD WAR II STRENGTHENED LINKS BETWEEN DEFENCE AND CIVILIAN SCIENCE, AS WELL AS REINFORCING RESEARCH LINKS BETWEEN AUSTRALIA AND BRITAIN

>> THE CSIRO

THE CSIRO, FROM ITS BEGINNINGS IN WORLD WAR I, HAS BEEN A CENTRE OF AUSTRALIAN SCIENCE RESEARCH EXCELLENCE

World War I provided an impetus for scientific research in Australia which coincided with and added to the dominance of science already emerging in the 20th century. In 1917–18 the Institute of Science and Industry was established. Its first purpose: 'the initiation and carrying out of scientific researches in connexion with, or for the promotion of, primary or secondary industries in the Commonwealth' (Rees ALG [1987] Ian William Wark 1899–1985, *Historical Records of Australian Science* 6(4)). This aim reflected the strong agricultural and pastoral basis of Australia's economy at the time, as did the work of agricultural scientists such as Farrer.

In 1926 the Institute became the Council for Scientific and Industrial Research (CSIR), and in 1949 its name changed again, to the Commonwealth Scientific and Industrial Research Organisation (CSIRO). The CSIR was initially concerned with research associated with Australia's primary industries, which then earned almost all of Australia's foreign income.

The first executive officer of CSIR—and later champion of the CSIRO—was parasitologist and veterinary scientist Ian Clunies Ross (1899–1959). Clunies Ross worked on parasites of domestic animals before becoming a science administrator. He headed the organisation during the introduction of the myxoma virus into Australia (Clunies Ross AI [1977] Ian Clunies Ross 1899–1959, *Records of the Australian Academy of Science* 3(3–4)).

The CSIRO was not only involved in biological research. One of first computers in world, CSIRAC, was designed and built in Australia in the late 1940s by Maston Beard and Trevor Pearcey, at CSIRO's Radiophysics Laboratory. This early Australian initiative was not followed up. The CSIRO was also home to organic chemist John Cornforth (1917–), who in 1975, with Vladimir Prelov, won the Nobel Prize for Chemistry, awarded for work on the structure of enzymes. Some of CSIRO's scientists produced technical equipment now used worldwide—Alan Walsh (1916–98), for example, invented the atomic absorption spectrophotometer developed by CSIRO's Division of Chemical Physics, and patented in 1954 (Golden Press 1986: 345).

Ian Wark (1899–1985) was another distinguished CSIR scientist. He joined the new Division of Industrial Chemistry in 1939 and became its chief. Wark studied overseas after his initial degrees at the Univeristy of Melbourne, but his major contribution came from research carried out when he returned to Australia—on conditions affecting the deposition of zinc ('Wark's rule'). With AB Cox, he clarified the role of reagents and the optimal conditions needed for mineral flotation, using methods which were later universally applied. His book *Principles of flotation* (Australasian Institute of Mining and Metallurgy, 1938) was widely read overseas, and used both in experiments and in plant processing.

in the world to observe the great comet of 1861—'one of the finest comets on record'—and 1881 (*Australian Dictionary of Biography* [1981], vol 6: 251). Tebbutt's astronomical observations have been commemorated by the naming of a lunar crater after him, an honour he shares with Roger Bacon, Leo Tolstoy and the mythical Roman messenger Mercury.

Among other 19th century Australian astronomers was Robert Ellery (1827–1908), who trained as a surgeon in England and came to Melbourne in 1852. Ellery founded the Williamstown Observatory and became its director and Victoria's government astronomer, positions he held for more than 40 years. The observatory moved in the 1860s to the Melbourne Domain, and in 1868 acquired the 'Great Melbourne Telescope', the largest in the world. The observatory became known around the world, and Australia's favourable latitude continued to make it a desirable base for optical and later radiophysics observations.

Not all important Australian scientific research has been based in and supported by institutions. Lawrence Hargrave (1850–1915) came to Australia from Greenwich, England in 1865, and trained as engineer. Though he worked in the late 1870s and early 1880s as an astronomical observer at the Sydney Observatory, in 1883 he left the observatory and worked privately on human flight and flight engineering. Hargrave's inventions included flapping wing 'orthopters'—aeroplanes which flew like birds by flapping their wings—and the box kite. He also designed and built engines to power flight. In 1884 he made his first flight, using the

PICTURE: AUSTRALIAN ACADEMY OF SCIENCE

> **ABOVE:** Sir Ian Wark (1899–1985)

tension of rubber bands; later he used compressed air for power. He received little local support. 'The people of Sydney who can speak of my work without a smile,' Hargrave wrote in 1892, 'are very scarce' (Moyal 1986: 175–76). But early US and European aeroplanes used the aeronautical principles of his box kite, and one famous early aviator, Gabriel Voisin, named his commercial aeroplane the 'Hargrave' (Australian Institute of Political Science 2000).

Although Australia's contributions to aeronautical engineering never quite took off, higher research into space proved more successful. From the late 1940s until the 1960s Australia led the world in radio astronomy research. As in many other fields throughout history, military needs and scientific advancement were closely entwined. Radio astronomy began in the United States in the 1930s with the work of Karl Jansky, who first detected radio waves from space, and Grote Reber, who built the first radio telescope. During World War II, radar technology was the focus of war science research in many countries, including Britain, Australia, Canada, the United States, Germany and Japan.

Australia's prominence in the field resulted in part from the government's decision at the beginning of the war to support radar research, and from the subsequent research done by small teams in the CSIR/O and within the University of Sydney (Bolton 1990). But it also has origins in Australia's colonial history. In 1939, several Commonwealth countries were secretly invited to send representatives to Britain to share her knowledge of radar (Robertson P [1992] *Beyond southern skies: radio astronomy and the Parkes telescope*: 21). It was a mutually beneficial arrangement. David Martyn went from Australia, returning with radar components, blueprints and research reports. Martyn, David Rivett and John Madsen set up the CSIR's innocuously named but strategically significant Radiophysics Laboratory, which initially worked, by agreement with Britain, on a particular area of radar research, but later became more independent in its direction (Robertson 1992: 21).

By the end of World War II the Radiophysics Laboratory was among the bigger CSIR divisions. In the United States and in England, but not in Australia, most staff at the radar laboratories (chiefly the US Radiation Laboratory and the Telecommunications Research Establishment in the UK) returned to their parent institutions. Staff of the Radiophysics Laboratory, whose work had shifted in emphasis gradually during the war years from an applied, weapons and defence-related focus to more basic research, carried the momentum of the war's radar research forward into radio astronomy research. In the 1950s there were research sites scattered around Sydney and throughout the state. Research conducted at these labs encompassed not only radar but radio astronomy, cloud physics and rainmaking.

Australia's radio astronomy research had two main arms by the early 1950s: solar/lunar, and cosmic studies. Solar and lunar radiation studies led to some interesting results. The moon's surface temperature varies by about 80°C over the course of its cycle. In 1948, attempting to explain the lag between the visible stages of the moon and the maximum, radiophysicists Jack Piddington and Harry C Minnett from the Radiophysics Laboratory concluded that radio emissions came from a layer under the moon's surface, and further deduced that its surface was 'porous rock and gravel covered by a top layer of fine dust with an average thickness of 2 cm' (Robertson 1992: 59), a conclusion confirmed by Apollo XI in 1969. Also in 1948, members of the Radiophysics Laboratory named the first three

EUROPEAN OCCUPATION OF AUSTRALIA CAME AT A TIME OF INTENSE INTEREST IN THE HEAVENS AND RAPID ADVANCES IN OPTICAL TECHNOLOGY. AUSTRALIA OFFERED A VIEW OF A DIFFERENT SKY

> **ABOVE:** The 'Bailey Boys'—Number 3 Radio Physics Training School, 1943

extraterrestrial radiowave sources to be identified with visible sources: supernova remnant Taurus A and the galaxies Virgo A and Centaurus A. They used radio interferometers (radio telescopes which measure interference to radiowaves) at Dover Heights in Sydney. Continuing the radio astronomy work, Dick McGee and John Bolton made a finding which caused US astronomer Walter Baade at Palomar Observatory to respond that 'Frankly, I jumped out of my chair the moment I saw what it meant.' The two astronomers had found the central point of the Milky Way, and they named this nucleus of our galaxy 'Sagittarius A' (Robertson 1992: 57).

Until the 1950s, science in Australia generally fitted the 'little science' mould, with low budgets, small equipment needs, and low operating costs. 'Big science'—complex projects such as manned space flights, undertaken by large teams and requiring extensive funds—was beyond the reach of individual universities and smaller institutions to fund and coordinate, and almost beyond the CSIRO. Its 64 m (210 ft) radio telescope, built 20 km north of Parkes, New South Wales, was a big science project, and could not have been built without US philanthropic fund money secured by the energetic EG ('Taffy') Bowen (1911–1991).

In the mid-1050s Bowen, chief of the Radiophysics Laboratory, used his large network of colleagues in the United States, and secured pledges of US$250,000 each from the Carnegie Foundation and the Rockefeller Foundation to develop a very large southern hemisphere radio telescope, in Australia. Without such substantial external support, Prime Minister RG Menzies was not willing to commit government funds to its design and building (Robertson 1992: 122).

AUSTRALIAN SCIENCE

Unlike the United States, Australia lacked a tradition of wealthy philanthropic trusts; these US foundations were prepared to contribute to a vast telescope in a different hemisphere which would have little immediate relevance to their own country.

Radio telescopes use the longer radio wavelengths which are transmitted through the Earth's atmosphere from space. There are two main ways of setting up such telescopes: as a large single telescope, or as a linked array. At the time of development, some of the Australian radio astronomers—including Bowen—favoured a single telescope. Others, including Joseph Lade Pawsey (1908–62), who had led the radio astronomy group under

During the World War II, the observatory at Mt Stromlo (Australian Capital Territory) and other academic institutions and government laboratories grew rapidly to undertake optical munitions research and production coordinated by the Optical Munitions (later the Scientific Instruments and Optical) Panel. Before World War II, optical glass had been manufactured only in Britain, Germany, France, the United States, Italy and Japan. During the war, Australia rapidly developed an industry in glass, lenses and prisms, and mirrors, as well as the machinery required for manufacturing these, and the ability to manufacture telescopes themselves. Australia supplied optical glass to the United States, New Zealand and India, as well as

UNTIL THE 1950S, SCIENCE IN AUSTRALIA GENERALLY FITTED THE 'LITTLE SCIENCE' MOULD, WITH LOW BUDGETS, SMALL EQUIPMENT NEEDS, AND LOW OPERATING COSTS

Bowen, pushed for an array. Bowen's preference was built. Later, the arrays favoured by Pawsey were built elsewhere in Australia.

The Parkes radio telescope was commissioned in 1961. At the opening, Bowen described the telescope as 'an instrument which is designed to extend our vision and understanding of the mysteries of outer space'. Ironically, its opening marked the end of what Robertson (1992: ix) calls the golden era of Australian radio astronomy, though the telescope and other Australian sites supported some of the most exciting outer space events in history—manned and unmanned space flights such as the Apollo moon landings and Voyager—as well as the discovery of quasars in the early 1960s.

domestically (Mellor DP [1958] 'Optical munitions', in *Australia in the War of 1939–1945*, series 4, vol 5: 246–281). 'At the end of the war, the Commonwealth Solar Observatory [Mt Stromlo], under Richard van der Riet Woolley (1906–86), had a first-class workshop and a spirit of working with optics that helped to establish their post-war work and reputation in optical astronomy' (Bolton 1990).

Woolley returned to England in 1956 as Astronomer Royal. In 1957, Mt Stromlo became part of the ANU. Woolley advocated and was instrumental in the building of the Anglo-Australian Telescope (AAT) at Siding Spring Mountain in the Warrumbungle Mountains (New South Wales), a type and size of telescope

PICTURE: AUSTRALIAN ACADEMY OF SCIENCE

> **ABOVE:** EG 'Taffy' Bowen (1911–91)

which he had first proposed while at Mt Stromlo. It began operating in 1975.

The 150-inch AAT was, when it was built, the largest optical telescope in the southern hemisphere, and is still one of the major facilities in the southern hemisphere. Siding Spring and Mt Stromlo are now combined observatories run by the ANU's Research School of Astronomy and Astrophysics.

At 12.56 pm on 21 July 1969, Honeysuckle Creek tracking station in the Australian Capital Territory transmitted television coverage of astronaut Neil Armstrong's first step on the moon, in what Apollo XI flight director Clifford E Charlesworth described as 'the greatest television spectacular of all time'. Nine minutes after the walk began, the moon came into view of Parkes, whose larger dish was then used to transmit its pictures to the United States and the world. Few who saw the moon landing have forgotten what they were doing on that day. The station was one of three NASA-directed Australian stations (Honeysuckle Creek and Tidbinbilla in the hills of the Australian Capital Territory, and Carnarvon, in Western Australia) covering the Apollo missions, including the manned Apollo XI space flight, and sending data to the flight's mission control centre in Houston, United States.

There is a pleasing symmetry in the astronomical events marking the span of time from Cook's Transit of Venus expedition to the lunar landing covered by the tiny station at Honeysuckle Creek.

>> ACKNOWLEDGEMENTS

This chapter relies heavily on the information gathered by Tim Sherratt and others into the Academy of Science's Bright Sparcs database, including articles from the Historical Records of Australian Science, and on the volumes of the Australian Dictionary of Biography. I am greatly indebted to Ian Coates, Peter Robertson and John Saxon for comments on drafts of the chapter.

PART TWO
EXPANDING THE FRONTIERS

AUSTRALIAN SCIENCE

INTRODUCTION

TO TRANSFORM INNOVATION INTO TANGIBLE OUTCOMES, THE GOVERNMENT INTENDS TO PROVIDE SUPPORT FOR THE SUCCESSFUL COMMERCIALISATION OF R&D, IDEAS AND INTELLECTUAL PROPERTY THROUGH POLICY INITIATIVES AND ADDITIONAL INFRASTRUCTURE FUNDING, REPORTS KEVIN PYLE

The Government's five-year *Backing Australia's Ability* plan is designed to address these issues by strengthening the country's innovation structure and supporting a full range of activities from pure research through to commercialisation. David Miles, Chair of the Innovation Summit Implementation Group, which advised the Government, said: 'To become internationally competitive we must be able to translate our ideas into tradeable products, processes and services. Ideas are the lifeblood of innovation. With commercialisation Australia will be an enviable innovation nation.'

A number of public and private organisations are taking a leading role in ensuring Australia's success in the commercialisation of R&D. They are doing this by addressing a number of key elements that have been determined to be essential for an effective innovation system, namely by creating strong links between industry and research providers; providing training, infrastructure and support; and creating a culture of innovation through education, awareness and entrepreneurship.

Cooperative Research Centres (CRCs) were established by the Government to forge links between the research capacities of universities and

EXPANDING THE FRONTIERS

public institutions with the needs of business. CRCs bring together researchers from universities, the CSIRO and other government laboratories, and private industry to foster collaboration of top researchers, which leads to effective and efficient research and research training. The critical mass achieved by CRCs allows them to make better use of resources—through the sharing of equipment and facilities—and, consequently, to work on long-term research and development projects of substantial size and quality.

Australian Technology Park Innovations (ATPI) is another organisation making a big difference in its efforts to help fast-track commercialisation. ATPI provides facilities and accommodation for R&D companies at the Australian Technology Park, bringing together university researchers, CRCs, spin-off companies and major players. Again, one of the primary goals of ATPI is to create links between researchers and industry, and to create a culture of innovation and support through the presence of its members. ATPI also has a number of education programs designed to provide essential support for companies that possess intellectual property, but lack the skills necessary to see their products through the prototype and development stages.

The ATPI's incubator program is a measure of the organisation's success. Nearly 50 companies have completed the two-year program and 37 companies, representing a range of disciplines, are currently enrolled in the program. The program not only helps incubatees in developing competencies, it also links them with professional services and financial underpinning, which is crucial to their growth.

MAKING THE MOST OF R&D INCENTIVES AND CONCESSIONS

MICHAEL JOHNSON & ASSOCIATES (MJ&A) is Australia's leading R&D focused support consultancy. It works closely with Australian businesses of all sizes in all industries to maximise the business benefits they realise from innovation. The firm's expertise is in government incentives for R&D, including tax concessions and grants.

Australian companies spend billions on research, development and commercialisation each year. Most know scientific and technological innovation is essential to get ahead of the competition and stay there.

Government also knows innovation is essential and spends millions of dollars on incentives and support for companies that invest in science and innovation. However, many companies fail to take full advantage of these programs leaving money on the table.

MJ&A offers an integrated portfolio of services ranging from high-level assistance with innovation planning and management to the nitty gritty of preparing compliance documents for yearly R&D tax concessions.

>> www.mjassoc.com.au
and turn to page 178 for directory details

AUSTRALIAN SCIENCE

> **ABOVE:** University of Melbourne scientists have developed a now commercialised way to make nanometre-sized metal particles known as 'quantum dots' with optical, electrical and magnetic properties that make them big news for suppliers of fluorescent inks, paints, phosphors and biolabels.

A HISTORY OF RESEARCH EXCELLENCE

SCIENTISTS AT THE UNIVERSITY OF MELBOURNE are world leaders in many fields. Gene research by Sir Macfarlane Burnett in the Department of Medical Biology at the Walter and Eliza Hall Institute won him a Nobel Prize in 1960. Subsequent research there unravelled the DNA sequencing of proteins.

Plant gene research led to cotton varieties containing a novel gene that is a natural insecticide and fast-track gene-level design and evaluation of wood cultivars for the timber industry.

Health-related research achievements include the bionic ear, an anti-tooth decay food and toothpaste additive from milk, and the discovery of genetic pathways in the malaria parasite making it vulnerable to new drugs and vaccines against animal parasite diseases.

Other industry-linked research has shown how to control small particles in suspension, produced a commercial process for extracting high-value materials from waste biomass and developed fast, accurate satellite mapping technology.

A land mine and parcel bomb detector that detects nitrogen in explosives and an environmentally clean hydrogen-assisted car engine fuel system are among new advances with major potential.

>> www.unimelb.edu.au
and turn to page 182 for directory details

Professor Steve Bakoss, CEO, says ATPI provides 'a framework of advice and mentoring that accelerates their development so they can do in two years what they might do in their research laboratories in five years, with a prime focus on developing competence and confidence in business practice and extending their potential for creative research outcomes.'

The creation of a culture of innovation is another key goal being achieved through a series of special courses. 'ATPI has a major role in sensitising bright young people to the opportunities of becoming an entrepreneur as opposed to becoming a "high-flying" professional. To this end we have a business/entrepreneurial program offered to the universities here in a collaborative sense,' Steve Bakoss says. The course aims to change people's perceptions and encourage the pursuit of careers in R&D—to teach students that they can be employers or creators, rather than employees.

Steve Bakoss believes the support of innovation as distinct from fundamental research is an important and nationally significant issue and that the Government's latest policy initiatives are very positive. 'It seems the Government is increasingly mobilised, they are going to the right people and formulating policy. The fact that both sides of politics say very similar things is also encouraging.'

The collaboration between research organisations and industry, coupled with Government support, has resulted in some spectacular results for Australia. The following chapters demonstrate the research efforts of some of Australia's dynamic success stories.

EXPANDING THE FRONTIERS

PROMOTING SCIENTIFIC AND TECHNOLOGICAL INNOVATION

AUSTRALIAN TECHNOLOGY PARK INNOVATION (ATPI) is a fully owned company representing the interests of Australia's four leading universities: University of Technology Sydney, The University of New South Wales, The University of Sydney and The Australian National University.

ATPI's primary function is the promotion of scientific and technological innovation through the management of the National Innovation Centre, the International Business Centre, the Business Incubator Program, the Advanced Manufacturing Centre and a number of research, educational and development initiatives.

Australian Technology Park is one of the most highly respected business and technology parks in Australia, and represents the largest collective of participants anywhere in Australia. There are more than 70 members of the ATPI community, including research and development initiatives such as the Redfern Photonics Group, multinational companies and sponsors such as Deloittes Touche Tohmatsu and Freehills.

ATPI provides facilities, accommodation and a critical mass that allows research and development (R&D) companies to collaborate with universities and industries in a real-world setting.

ATPI also plays a crucial role in the initiating stages of R&D commercialisation. Its Business Incubator Program offers a framework of advice and mentoring, as well as links to professional services and financial underpinning. Since its inception four and half years ago, more than 50 participants have taken part in the two-year program. There are currently 37 companies in the program representing a diverse range of disciplines.

By bringing these companies together in one location, ATPI has created an environment that fosters synergy between participants. Members are able to share, and be encouraged by, the successes of other participants and often, organisations coalesce, sharing resources and knowledge.

In research areas such as IT, where the risks can be huge, ATPI has had significant successes. Nearly 80 per cent of participants accomplish what they set out to do. Recent success stories include Elcom Technology, an innovator of business-driven, e-commerce solutions; Indian Pacific Communications, which was bought by listed company Open Telecommunications Ltd; SmartContainers Limited a company which is developing innovative technologies to monitor and track refrigerated shipping containers throughout the world; and Pure Commerce, winner of the award for 'Most Innovative E-Commerce or Internet Start-up' at the Australian Technology Awards 2000.

>> www.atp.com.au
and turn to page 172 for directory details

Dispensing suite for intravenous plasma products in CSL's plasma products facility at Broadmeadows in Melbourne

HEALTH AND MEDICAL

AUSTRALIAN SCIENCE

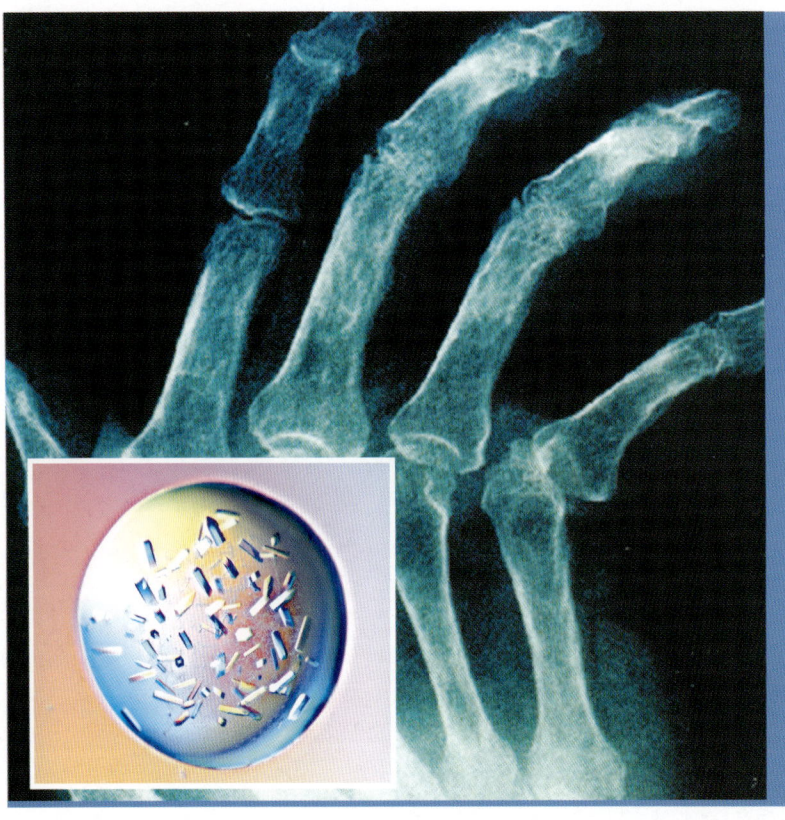

> **ABOVE:** Crystals of human Fc Receptor (FcμRIIa) produced by Dr. Maree Powell after 7 years. These were used to determine the Fc Receptor structure, which forms the basis of a new approach to design new drugs to treat autoimmune diseases such as rheumatoid arthritis (x-ray pictured above).

NEW WAYS OF DETECTING AND ERADICATING DISEASE

THE AUSTIN RESEARCH INSTITUTE (ARI) is a world leader in developing vaccines for cancer, improving organ transplantation and understanding the nature of inflammatory diseases. An ARI vaccine program in cancer, which targets the mannose receptor, is now in clinical trial. The ARI also now has a major commitment to develop vaccines for infectious diseases.

ARI was the first group to crystallise an Fc Receptor for IgG and to determine its structure. Inhibitory drugs to this combination have been synthesised and are being studied for the treatment of disease particularly rheumatoid arthritis and bleeding diseases.

ARI also specialises in the transplantation of animal organs to humans, demonstrated the role of anti-Gal antibodies in hyperacute rejection by isolating the pig Galactosyl-transferase gene. The ARI has designed methods of down regulating the expression of this gene. Complement inhibitors to enable prolonged xenograft survival have been defined and there is an active program in pig-to-primate transplantation in place.

>> www.ari.unimelb.edu.au
and turn to page 171 for directory details

When Australian Professor Peter Doherty shared the 1996 Nobel Prize for Medicine, he was maintaining the nation's reputation for excellence in medical research. That high standard is set to carry on into the 21st century. In this chapter we look at some of the many key players around Australia.

In fact, right now cutting edge research is under way at the very same institution where Peter Doherty and a young scientist from overseas, Rolf Zinkernagel, made their award-winning discovery.

In 1973, the pair were conducting experiments with mice at the Australian National University's John Curtin School of Medical Research (JCSMR). By chance, they stumbled onto the solution to a longstanding scientific mystery: how does the body's disease-fighting immune system detect viruses which are hidden inside infected cells?

The answer is that specialised immune system cells, T cells, recognise a molecule on the infected cell's surface which is altered when a virus sneaks into it. This discovery has helped the development of potential prototype vaccines against cancers and many deadly viruses.

One of those killer viruses is Human Immunodeficiency Virus (HIV), the virus which causes AIDS. At the JCSMR, Professor Peter Gage and his team have found a compound that stops HIV replicating in a certain type of cell culture. The compound stops the spread of HIV by blocking ion channels. Ion channels are large molecules that act like gateways, allowing small particles to pass in and out through the protective membrane of a cell. When the ion channel is blocked, the virus cannot escape, or 'bud off',

HEALTH AND MEDICAL

from the infected cell. This work, and related projects, is supported by the team's new biotechnology firm, Biotron.

At the University of Technology, Sydney (UTS), Associate Professor Donald Martin is also studying ion channels. He has discovered that blocking ion channels on macrophages reduces the stickiness of these large white blood cells. The finding is significant because macrophages stick to plaque inside blood vessels. When the build-up gets too bulky, the blood vessels clog. This condition, atherosclerosis, can be life-threatening. Martin believes his research could lead to a clot-busting drug to treat the illness.

Ion channels have also emerged as a possible cause of epilepsy, a condition which causes convulsions and unconsciousness. Professor Sam Berkovic, of the Epilepsy Research Institute, and his colleagues have identified two of the four known 'epilepsy genes'. Both these genes code for (tell the body to build) ion channels that allow sodium and potassium to pass through cell membranes.

The brain itself is also a primary research area for POWMRI scientists. The Institute houses one of only five 'brain banks' in the world. Stored in the bank are over 400 brains, which are used in the study of neurodegenerative diseases such as Alzheimer's and Parkinson's Disease.

Mysteries behind many other mental disorders are also being solved, thanks to Australian researchers. Professor Grant Sutherland—former president of the prestigious Human Genome Project Organisation (HUGO)—made a major breakthrough in the understanding of fragile X syndrome, an inherited

DESIGN, MANAGEMENT AND ANALYSIS OF CLINICAL RESEARCH TRIALS

ONE OF ELI LILLY Australia's major research initiatives is the establishment of the Clinical Outcomes and Research Institute as a regional centre of excellence for clinical and outcomes research trial design, management, analysis.

The Institute has been designed to achieve its overall goal: to target, and provide solutions for, many of the world's most urgent, unmet medical needs.

It will research and develop new medicines for the treatment of cancers, cardiovascular diseases, endocrine diseases such as diabetes and osteoporosis, infectious diseases and neurological disorders such as depression and schizophrenia.

Links are being formed via the Institute with Australian laboratories, hospitals, contract research organisations, strategic research partners, research alliances and academia to provide networks essential for fulfilling the Institute's research goals.

The Institute will also develop further strategic alliances with other Australian research institutions, increase Eli Lilly Australia's involvement in all stages of clinical development and increase the company's overall level and scope of research in Australia.

>> **www.lilly.com.au**
and turn to page 175 for directory details

AUSTRALIAN SCIENCE

MICROARRAYING GENES ONTO GLASS SLIDES OR CHIPS

THE GARVAN INSTITUTE OF MEDICAL RESEARCH was first in Australia to establish an oligonucleotide gene microarray facility fitted with a sophisticated gene chip analyser. As various genomics technologies transform the face of medical research the ability to analyse thousands of genes at a time is a feat analogous to switching from riding a bicycle to flying by jet.

The Garvan's cancer and arthritis programs have already identified several genes that are turned 'on' or 'off' in diseased tissue. These sets of genes are important targets for developing new approaches to the treatment of cancer and arthritis.

The Garvan strives to make significant contributions to medical research that change the directions of science and medicine and have major impacts on human health. A combination of fundamental science with strong clinical applications have made it one of the world's pre-eminent medical research institutes.

>> **www.garvan.org.au**
and turn to page 175 for directory details

cause of mental retardation. Sutherland, of the University of Adelaide and the Women's and Children's Hospital, discovered a previously unknown way of passing on defective genes. The finding led to a genetic test that can assess a couple's risk of having children with the syndrome.

Working at both the Mental Health Research Institute (MHRI) and the University of Melbourne, a team led by Professor Colin Masters is working on a treatment for Alzheimer's Disease. They hope to halt the brain damage which leads to problems with thinking and memory. With colleagues at Harvard University, Masters' group has conducted laboratory experiments with animals. The results reveal that a type of chemical called a chelator acts to reverse the toxic processes involved in damaging the brain. Clinical trials have begun.

Scientists with the MHRI are also tackling schizophrenia. Associate Professor Brian Dean's laboratory has discovered a complex molecule, a protein, that is involved in the speed at which brain cells, or neurones, exchange information. Dean predicts that the protein helps cause the delusional state experienced by people with schizophrenia. If so, it will be a prime target for drug designers, who could build better anti-psychotic medications with fewer of the unpleasant side effects associated with existing drugs.

Australian twins have led to a breakthrough in another mental illness: severe anxiety. At the Queensland Institute of Medical Research, Adjunct Professor Nicholas Martin has studied the personality traits of twins and siblings from 6,000 families. He has discovered that the same genes

HEALTH AND MEDICAL

PROMOTING THE UNIQUE ACHIEVEMENTS OF AUSTRALIA'S RESEARCH INSTITUTES

THE ASSOCIATION OF AUSTRALIAN MEDICAL RESEARCH INSTITUTES (AAMRI) was formed in 1993 to facilitate communication and collaboration amongst independent medical research Institutes throughout Australia.

Research projects carried out by members of AAMRI involve practically every aspect of human health and disease. Current projects range from cancer, AIDS and schizophrenia to Aboriginal and international health, fundamental laboratory research and the design, implementation and evaluation of practical disease prevention programs.

The 23 members of AAMRI form an internationally recognised component of Australian biomedical research and perform more than one quarter of all non-commercial biomedical research carried out in this country.

Member Institutes have a close working relationship with Australian and multinational pharmaceutical companies, Commonwealth research centres and research syndicates around the country. This network enables them to play a major part in the commercialisation of Australian biomedical discoveries.

AAMRI's overriding goal is to improve awareness of the unique achievements and needs of medical research Institutes amongst governments, industry and the public. It endeavours to ensure that governments, business and the public are aware of the vital contributions our country's medical research Institutes make to improving human health–both in Australia and around the world.

The Institute plays a major role in coordinating communication between research Institutes, improving scientific and administrative collaboration and helping communicate their collective views and interests to governments and the Australian public.

The Institute also works to create and maintain a fiscal and regulatory environment which enables Australian medical research to be carried out in an efficient and economical manner, and that the fruits of this work can be the practically applied.

The aggregate budget for AAMRI members is well over $100 million per year, and they employ more than 2000 scientists. They also contribute enormously to the scientific labour supply by training BSc Honours, Masters and PhD students and postdoctoral fellows.

>> morrellk@cryptic.rch.unimelb.edu.au
and turn to page 169 for directory details

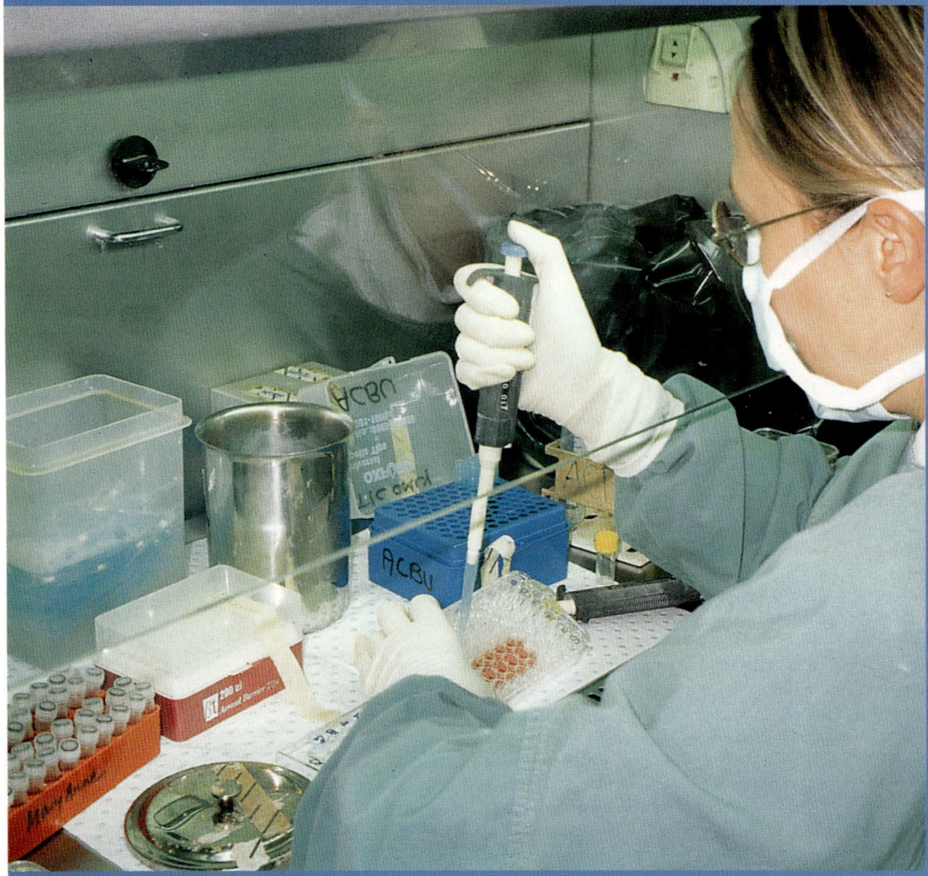

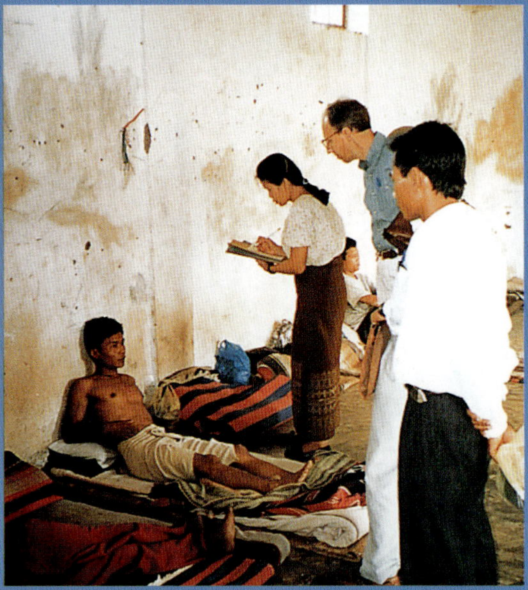

AUSTRALIAN SCIENCE

> **ABOVE:** HIV/AIDS community health project in South East Asia.
> **INSET:** HIV/AIDS researcher working on virus in a biohazard cabinet.

PUBLIC HEALTH RESEARCH AT WORK—FROM THE LABORATORY TO THE LOCAL CLINIC

THE MACFARLANE BURNET CENTRE FOR MEDICAL RESEARCH was founded in 1986 and is now Australia's premier virology and public health research institute.

The Centre is unique in Australia due to its involvement in laboratory, clinical, epidemiological and public health research as well as the design, implementation and evaluation of practical public health programs.

The Centre's laboratory work in collaboration with the New South Wales Red Cross has led to the characterisation of a unique strain of HIV which is much less pathogenic than regular strains. The new strain either does not cause AIDS in infected patients, or does so only very slowly, and is considered an important starting point for the development of a live attenuated HIV vaccine which might stem the current epidemic of HIV infection.

Their laboratories have also developed a test for the hepatitis E virus (HEV) infection—a common cause of hepatitis in developing countries. The new test has been utilised by the Centre in studies revealing that HEV infection is more prevalent in a number of Asia-Pacific countries than previously thought. The Centre is now developing public health programs to limit the spread of HEV and, at the same time, exploring the potential for an HEV vaccine.

The Centre attempts to improve the health of citizens of developing nations through the rational and rigorous assessment of public health issues and public health program outcomes.

One recent project was conducted in collaboration with the Program for Appropriate Technology in Health (PATH) based in Seattle in the USA. The goal was to immunise newborns on the island of Lombok in Indonesia against hepatitis B virus infection. This study showed it was possible to reduce the rate of active hepatitis B infection by more than 75 per cent in a resource-poor setting via a simple and inexpensive hepatitis B immunisation regimen.

At the end of 2001, The Macfarlane Burnet Centre for Medical Research will move into a 5,000 square metre purpose-built research facility within the Alfred Hospital Medical Research and Education Precinct in Prahran, Victoria.

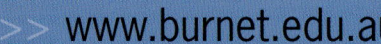
>> www.burnet.edu.au
and turn to page 177 for directory details

that predispose people to anxiety and depression contribute to 'neurotic' behaviour such as excessive worrying, fearfulness of social situations and mood swings. Right now, the British firm Gemini Genomics is working to turn the genes into a drug which will target those conditions directly.

Type 1, or insulin-dependent diabetes, is a long-term, often debilitating disease. It is caused when the pancreas does not produce enough insulin, a hormone which regulates blood sugar levels. Symptoms of insulin deficiency include extreme hunger and thirst, and metabolic failure. Although people with diabetes can control their disorder with daily injections of insulin, they can still expect complications such as blindness, kidney failure, and even the loss of limbs. They are also at risk of a reduced life expectancy. All of this explains why scientists around the world are seeking a needle-free, more naturally regulated system for producing the right amount of insulin at the right time. Australian researchers are no exception.

At UTS, Dr Ann Simpson is using gene therapy to treat diabetic mice. The procedure involves inserting the human insulin gene into human liver cell lines and these were transplanted into diabetic mice. Liver cells are used because the mice pancreatic cells are defective or dead. So far, the livers of the treated mice are successfully producing insulin, without the need for injections. The little animals seem hale and hardy. Simpson predicts that the technology will be ready for trials with humans within the next few years.

Further down the track is the possibility that people with Type 1 diabetes could simply 'grow'

HEALTH AND MEDICAL

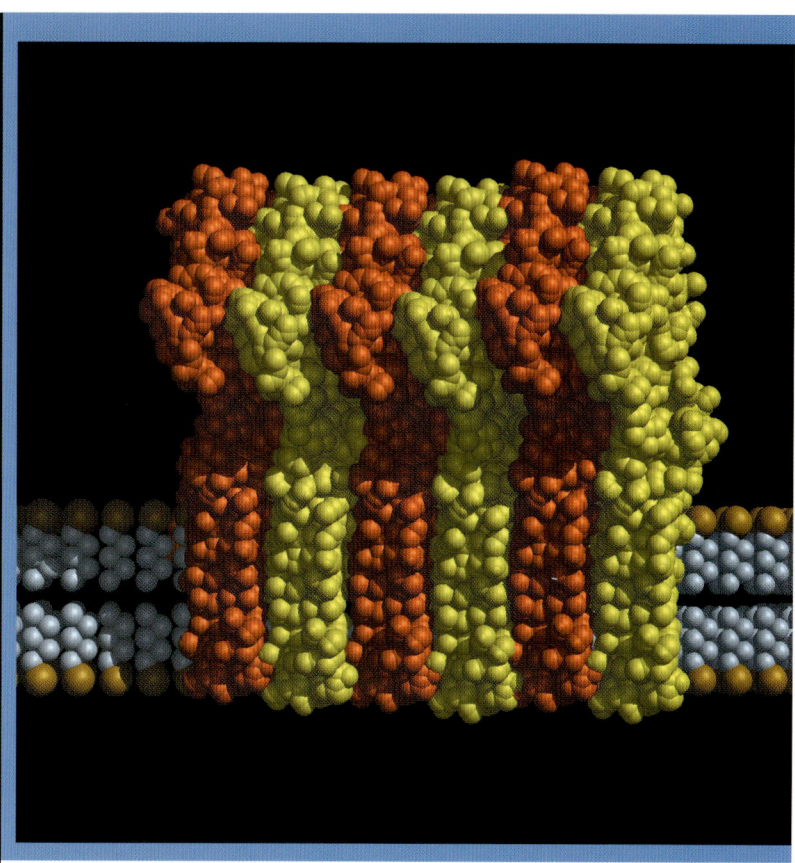

> **ABOVE:** Computer image of how the protein toxin perfringolysin is inserted into the cell membrane, disrupting cell function.

PROTEIN STRUCTURE PROVIDES DRUGS FOR CANCER AND HEART DISEASE

ST. VINCENT'S INSTITUTE OF MEDICAL RESEARCH has concentrated on the structure and shape of proteins since its foundation in 1957. It explores how normal amounts of proteins of specific amino acid sequence and shape are essential for life and how subtle changes can lead to diseases such as cancer and heart disease.

Current work on protein structure builds upon research by founding Director Pehr Edman who developed and automated the chemical method used to determine the amino acid sequence of proteins. The Institute has solved 20 protein structures during the last eight years using protein X-ray crystallography. Successes include determining the structure of toxins such as perfringolysin O—a member of a family of toxins responsible for diseases including pneumonia and gas gangrene.

The Institute's work on protein structure underpins its research into breast and prostate cancer, bone cell biology (osteoporosis) and blood pressure, and provides the basis for the design of drugs to prevent such diseases.

>> www.svimr.unimelb.edu.au

and turn to page 181 for directory details

AUSTRALIAN SCIENCE

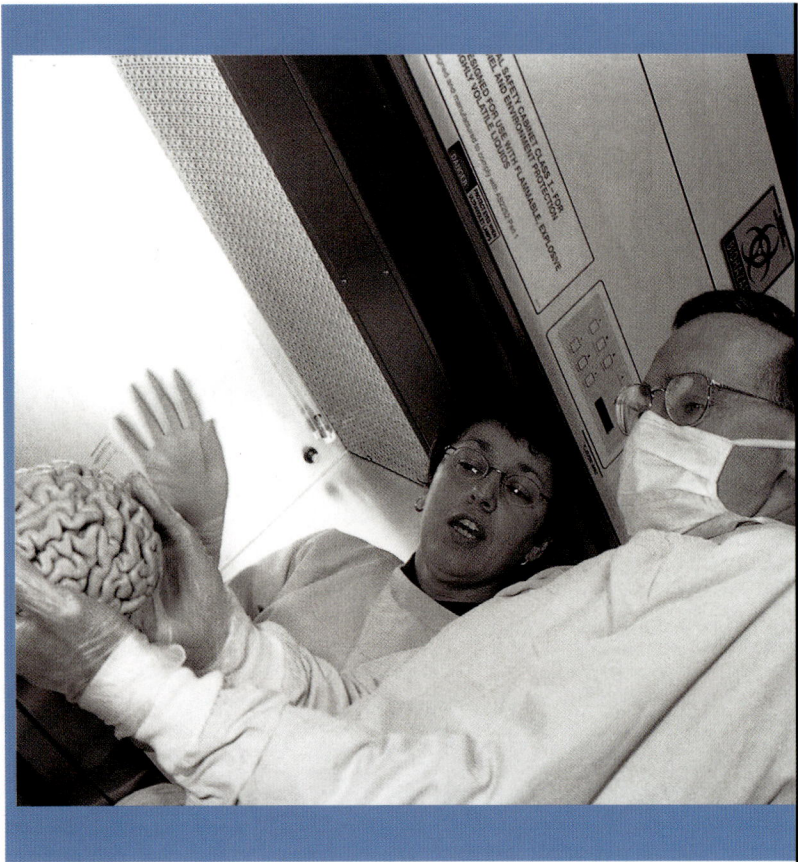

> **ABOVE:** Associate Professor Glenda Halliday and The Hon. Bob Carr, Premier of NSW, inspect the new facilities for research into Parkinson's disease.

PARKINSON'S DISEASE BREAKTHROUGH

Researchers at the PRINCE OF WALES MEDICAL RESEARCH INSTITUTE in Sydney are developing objective tests for Parkinson's Disease.

The Institute is Australia's leading centre for research on the functions and disorders of the brain and nervous system and has major research programs in place on several serious neurodegenerative disorders.

Their simple blood test for Parkinson's Disease is expected to identify sufferers before symptoms such as tremors, muscle stiffness and movement problems appear. An early warning system such as this is a crucial advance in treating the disease.

Parkinson's symptoms usually appear seven to ten years after the disease has started. However, by this time, around 70 per cent of the brain cells that control movement have already died.

Not only do existing therapies become less effective as the disease progresses, new drugs under development are aimed at preventing this cell death. If taken early enough they might prevent symptoms appearing and, thus, effectively prevent Parkinson's Disease.

>> **www.powmri.unsw.edu.au**
and turn to page 180 for directory details

a new insulin-producing pancreas. This is called therapeutic cloning, and it is one of the most exciting areas of medical research today.

The goal is to repair ageing or damaged bodies with real human tissue, grown especially for the task. The conditions that could be treated are virtually endless: anything from spinal cord damage, heart disease, Parkinson's Disease and liver damage to stroke. Anywhere where an organ or tissue is not doing its job, intervention may be possible.

Scientists around the world are exploring two ways of creating designer tissue. One is to grow the 'spare parts' in animals, such as pigs. These donor animals would be genetically altered so that the organ or tissue would not be rejected by the human recipient's immune system.

However, a more promising technique is to use human cells which have been tricked into growing the required part. But which cells to use?

The embryonic stem cell is the cell of choice. These so-called 'pluripotent' cells have the capacity to develop into any and every type of cell in our bodies. International teams have had early success, coaxing stem cells to turn into various sorts of cells, from muscle and liver to nerve cells.

The catch is that these stem cells come from embryos. Some people oppose the use of stem cells because the extraction of the cells destroys the embryo, even though it is only a few days old. Technically, this bundle of cells is called a blastocyst, because it has not begun to develop embryonic tissues and organs. A national committee is currently exploring these ethical issues.

HEALTH AND MEDICAL

A team led by Professor Alan Trounson and Dr Martin Pera at Monash University has sidestepped an Australian ban on using human embryonic stem cells by 'harvesting' the cells in Singapore. They were the first to turn the cells into nerve cells. The group has since joined with researchers from Singapore and Israel to find out more about how key cells—heart, gut, and nerve cells—work in the body.

This is important because the group, like other scientists, is now investigating methods of turning 'differentiated' adult cells—say nerve cells—into 'undifferentiated' stem cells. If stem cells can be 'made' from adult cells, there is no ethical problem associated with their use. Already, the Monash team has successfully turned adult mouse cells into stem cells.

Moreover, if such procedures can be developed, human therapeutic cloning would be possible: a patient's damaged tissue would be cloned by fusing one of his or her cells with a human egg with the nucleus removed. The egg would develop into an embryo, from which stem cells would be extracted. The stem cells would be grown into the target tissue.

The benefits are obvious: no ethical issues, and no possibility of immune rejection. And perhaps a cure for cancer.

Cancer research in Australia, from treatment advances to discoveries about the complex process by which the disease forms and spreads throughout the body, is gaining international recognition.

As the nation's only comprehensive and dedicated cancer research and clinical facility, the Peter MacCallum Cancer Institute (PMCI) is the

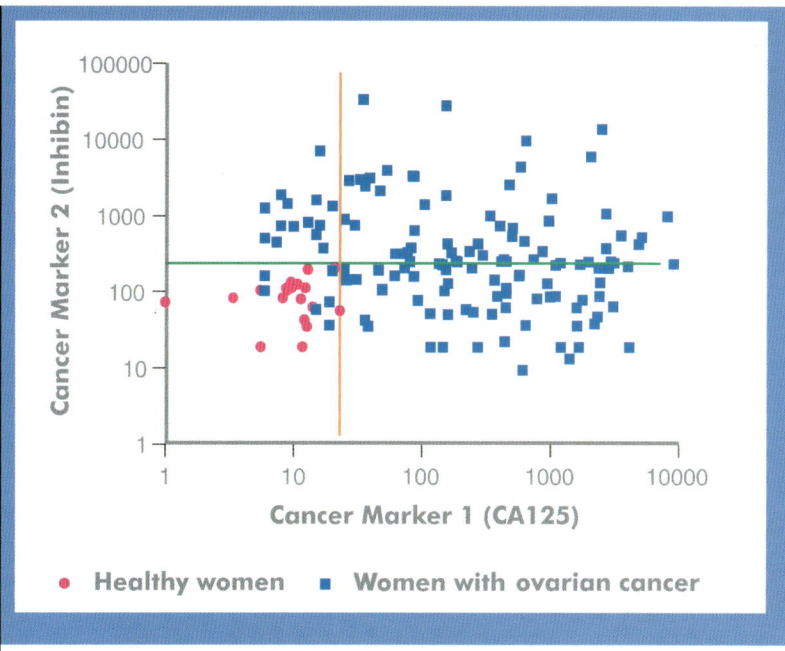

> **ABOVE:** This graph shows how the combination of a recognised cancer marker (CA125) with a second and new cancer marker, inhibin, is better able to distinguish between healthy women (red dots) and women with ovarian cancer (blue dots). When testing with CA125 only, we can see that all the women on the left hand side of the orange vertical bar test cancer free. But, by combining this test with Inhibin, about half of the women (those above the green line) that would have been presumed cancer free, show evidence of cancer. This has important implications for the successful and early diagnosis of ovarian cancers.

INHIBIN—A KEY TO OUR UNDERSTANDING OF REPRODUCTION

PRINCE HENRY'S INSTITUTE OF MEDICAL RESEARCH in Melbourne is responsible for a groundbreaking discovery in 1984, when scientists there identified a new hormone called inhibin. At the time, they didn't realise how significant their discovery would be.

Inhibin turns out to be one of the major hormones involved in fertility. Without knowing that inhibin is vital for the development of eggs and sperm, much of what medical experts now know about reproduction would still be a mystery.

Inhibin is also vital for the detection of some cancers. Inhibin falls to very low levels in women after menopause, while some ovarian cancers produce their own inhibin. Measuring inhibin levels can lead to early detection, diagnosis and treatment of these cancers, and accurate new tests are currently being developed.

Not all is known about inhibin's role in the body as yet, but ongoing research at the Institute will shed more light on the role of this important new hormone.

>> www.med.monash.edu.au/phimr
and turn to page 180 for directory details

AUSTRALIAN SCIENCE

> **ABOVE:** Scientists working in the Centenary Institute's state-of-the-art research laboratories located in the grounds of Royal Prince Alfred Hospital.
> **INSET:** The FACStar in the flow cytometry facility—a vital tool in the Institute's cell based research.

NEW APPROACHES FOR DIABETES AND ASTHMA

THE CENTENARY INSTITUTE OF CANCER MEDICINE AND CELL BIOLOGY is working on a potential vaccine for insulin dependent diabetes which could be ready for clinical testing within the next year. A new approach to the drug treatment of asthma has also been devised.

The Centenary Institute is, according to 1996 Nobel Prize winner, Professor Peter Doherty AC, Australia's 'premier immunology research centre'. It specialises in immunology and its application to the diagnosis, prevention and treatment of conditions ranging from cancer, arthritis and organ transplant rejection to insulin dependent diabetes, asthma and tuberculosis.

The Centenary Institute is working on the problem of why cancer cells survive so much better than normal cells and is looking at the genetic defects underlying leukaemia, breast cancer and colon cancer so that these can be corrected by gene therapy. Ultimately the Institute's cancer research program will focus on the development of preventative strategies in the form of vaccinations for high risk subjects.

>> **www.centenary.usyd.edu.au**
and turn to page 173 for directory details

flagship of Australian cancer science. Following the work of scientists such as Alan Trounson and Martin Pera, PMCI experts have established a working group to explore the use of stem cells in cancer therapy.

PMCI scientists are also trialling a new use of positron emission tomography (PET). PET is a non-invasive way of seeing, or imaging, how a body is functioning. The technology is often used to diagnose cancer. But Associate Professor Rodney Hicks believes it can also help 'stage', or assess, a patient's treatment. Often the only way to stage a cancer is by surgery. Results so far indicate that PET can indeed be used to replace other expensive or invasive staging procedures—and it has also been found to be more accurate.

Another PMCI team, headed by Dr Andrew Holloway, is applying the new 'gene chip' technology to the diagnosis of cancer. Gene chips are tiny arrays of known DNA fragments which have been fixed to a solid surface. They function as a biological database, against which samples from patients can be compared. Experts can now put markers for 5,000 genes on a chip. By the end of this year, they expect to increase that number to 40,000.

One important application of gene chip technology is the identification of the site of a patient's 'primary' tumour. This is where the cancer developed, before spreading to other parts of the body. No matter where it spreads, a tumour carries the genetic 'fingerprint' of its 'organ of origin', for instance the lymph nodes, liver, bone or lungs. If doctors know the source of a tumour, they can tailor a patient's therapy more

HEALTH AND MEDICAL

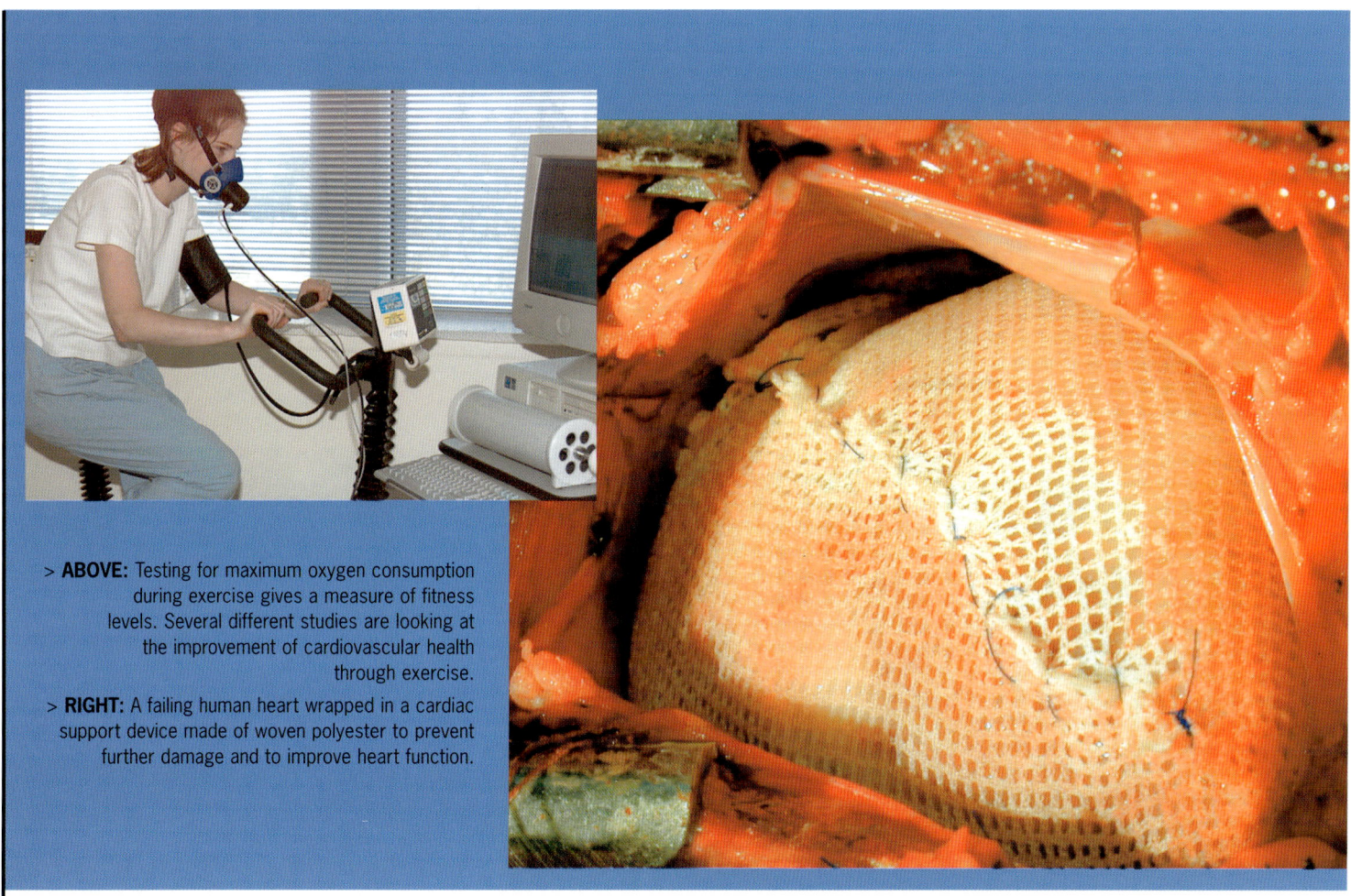

> **ABOVE:** Testing for maximum oxygen consumption during exercise gives a measure of fitness levels. Several different studies are looking at the improvement of cardiovascular health through exercise.
> **RIGHT:** A failing human heart wrapped in a cardiac support device made of woven polyester to prevent further damage and to improve heart function.

UNDERSTANDING THE CAUSES OF HYPERTENSION AND ATHEROSCLEROSIS

Research at the BAKER MEDICAL RESEARCH INSTITUTE (BMRI) covers cardiovascular diseases such as high blood pressure and atherosclerosis and investigates how disease states of the heart and blood vessels are influenced by the central nervous system.

The BMRI's aim is to understand what causes hypertension and atherosclerosis and apply this knowledge to the prevention and treatment of heart and blood vessel disease.

Expertise in basic and clinical research among BMRI scientists supports a 'benchtop-to-bedside' approach in which basic research findings are then applied to patient care. Among the contributions the BMRI researchers have made to the health and well-being of Australians are:

> The knowledge that regular, moderate exercise is good for cardiovascular health because it lowers blood pressure and improves the profile of blood lipids. An increase from moderate to heavy exercise produces no further benefit for heart disease.
> The discovery of a biochemical link between heart disease and acute mental stress responses, panic disorder and depressive illness. The BMRI's innovative research techniques now enable the scientific study of the mind/body relationship which is often used to explain unknown aspects of heart disease.
> A cardiac support device, made of special woven polyester, has been shown to halt the chain of damaging events that ends in heart failure.
> A BMRI knockout mouse model which is unable to respond to an important stress hormone demonstrated that stress, known to affect the heart and blood vessels, does not interfere with the developing thymus.
> People with heart failure have low levels of a nerve growth factor called 'NGF' which may explain why nerves to the failing heart are constantly switched on instead of responding only when increased blood flow is required.
> Restricting calories and increasing exercise, the most common non-drug treatments for hypertension, have been shown to act by inhibiting the sympathetic nervous system.

>> www.baker.edu.au
and turn to page 172 for directory details

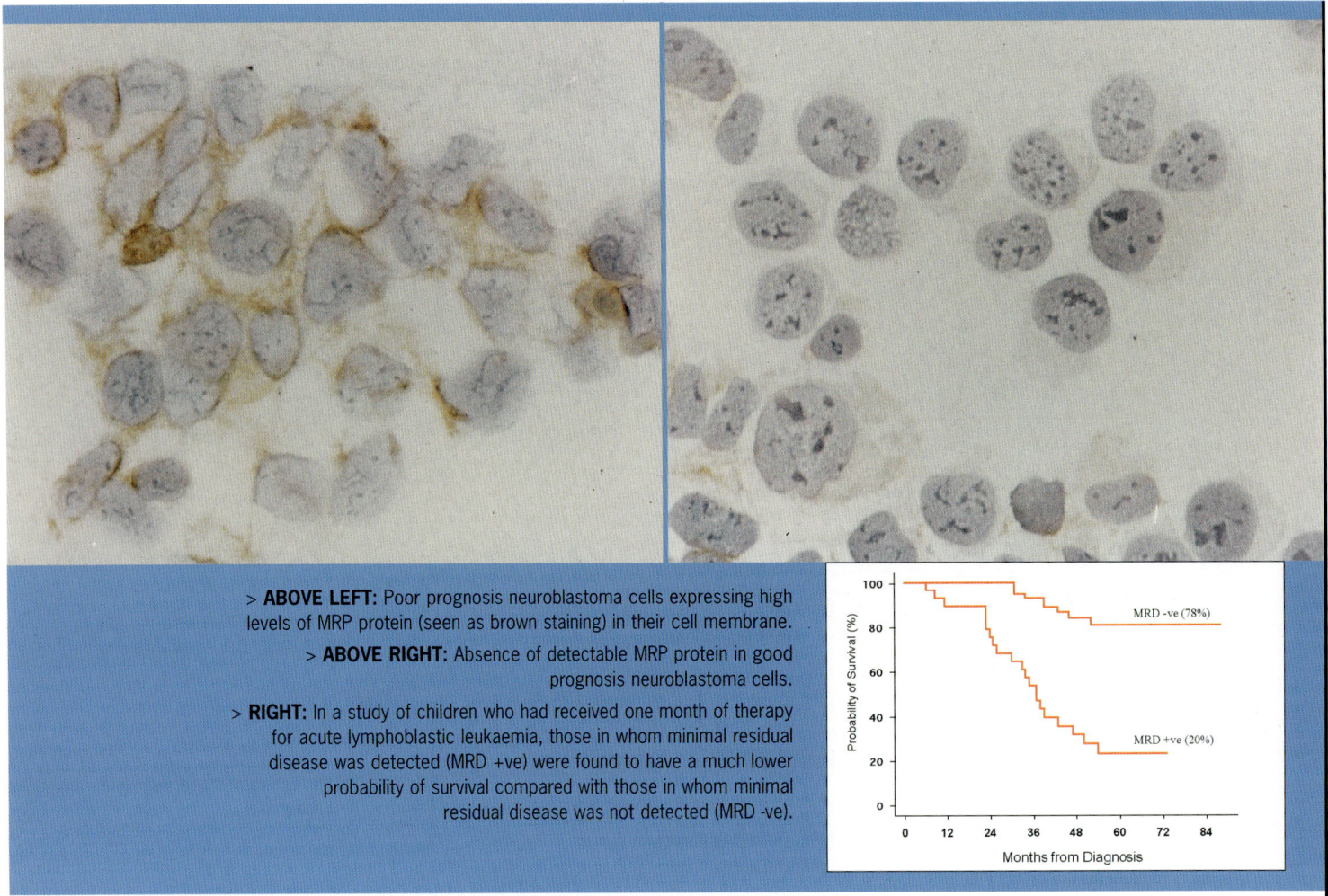

> **ABOVE LEFT:** Poor prognosis neuroblastoma cells expressing high levels of MRP protein (seen as brown staining) in their cell membrane.
> **ABOVE RIGHT:** Absence of detectable MRP protein in good prognosis neuroblastoma cells.
> **RIGHT:** In a study of children who had received one month of therapy for acute lymphoblastic leukaemia, those in whom minimal residual disease was detected (MRD +ve) were found to have a much lower probability of survival compared with those in whom minimal residual disease was not detected (MRD -ve).

RESEARCH FOCUSED ON CHILDHOOD CANCERS

CHILDREN'S CANCER INSTITUTE AUSTRALIA FOR MEDICAL RESEARCH has a vision to save the lives of all children with cancer and eliminate their suffering through world-class research.

The Institute is an independent medical research facility affiliated with Sydney Children's Hospital and the University of New South Wales. It is the only Institute in Australia devoted solely to research into the causes, prevention, treatment and cure of childhood cancer.

Recent success stories have included advances in understanding the most common solid tumour in young children – neuroblastoma. The Institute has shown that neuroblastoma patients whose tumours contain high levels of the multi-drug resistance-associated protein (*MRP*) gene are at far greater risk of relapse and death than those with low levels. This research, published in *The New England Journal of Medicine*, led to the award of the Seligson Prize in 1996 and the establishment of a large international study in collaboration with the Pediatric Oncology Group of America. The Institute is currently evaluating compounds that may reverse MRP-mediated cytotoxic drug resistance in neuroblastoma.

Other research has included the early detection of relapse in children with acute lymphoblastic leukaemia (ALL), the most common childhood malignancy. Over 95 per cent of children with ALL achieve complete remission after standard treatment, however, the disease returns in around one third of these children. These relapses are due to a small number of previously undetectable cancer cells that survive or escape treatment. The Institute has developed a new technique that can detect one cancer cell among a million normal cells and predict relapse over 12 months before any clinical evidence emerges. A rapid test for residual disease has been developed for routine clinical use and trials are planned for 2001 in which alternate treatment strategies will be implemented for children at high risk of relapse. This work was recently published in the *British Journal of Haematology*.

>> **www.ccia.org.au**
and turn to page 173 for directory details

HEALTH AND MEDICAL

effectively. Unfortunately, in 3 to 5 per cent of all new cases, doctors do not know whereabouts in the body the cancer started. These patients have a poor prognosis.

PMCI scientists are building gene chips that pinpoint a tumour's source by identifying the organ's telltale fingerprint. A preliminary trial, conducted with Stanford University researchers, suggests that the technology will work both quickly and accurately.

Gene chips are also being used at the PMCI to find the genes that cause the spread, or metastasis, of breast cancer. The aim is to better predict the future course of the disease for each patient. This too will lead to better treatment.

Thanks to gene chips and other advances in human genetics, John Curtin School of Medical Research scientists have gone high-tech with mice. A group headed by Dr Simon Easteal is building a mouse 'library'. Scientists can more easily study specific genes in mice than in people. They can use mice because almost all the genes in mice have a human counterpart that is only slightly different.

The library consists of frozen sperm and genetic material from mice strains which have been bred with a mutation on half of their chromosomes (the material carrying their genes). The mutations are 'deletions', known to cause a disease, such as diabetes or cancer, in the mice. The deletion library will help speed up the process of 'making' mice with the right genes for research. And that means research results will come more quickly.

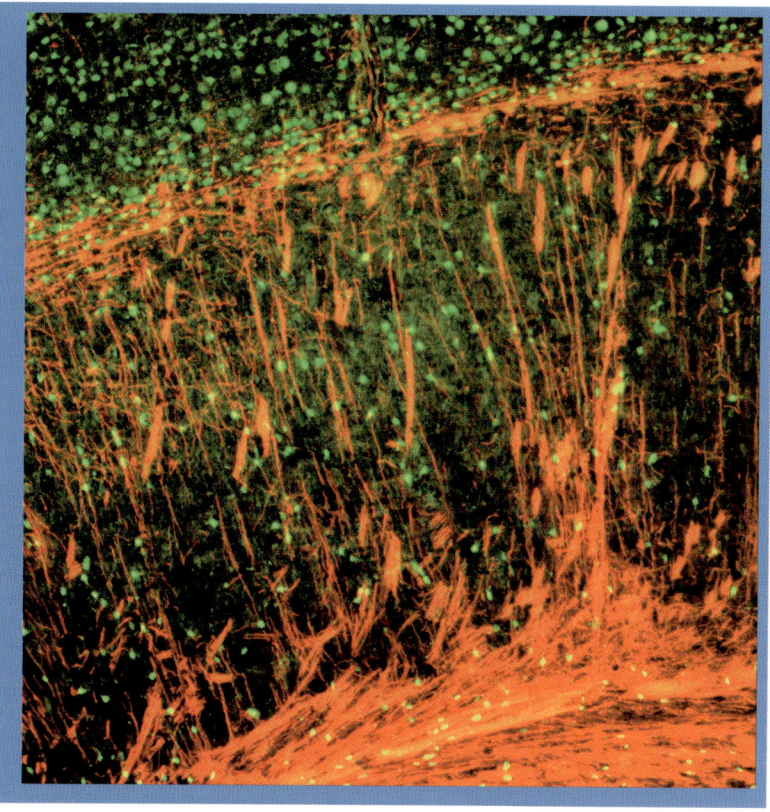

> **ABOVE:** CNPase staining (red) showing areas of myelination in the mouse brain.

UNDERSTANDING EPILEPSY AND SCHIZOPHRENIA

THE HOWARD FLOREY INSTITUTE'S neural development group wants to understand how the mammalian brain develops in order to shed light on how the evolution of the brain contributed to the uniqueness of humans and to understand the disease processes causing mental illnesses such as epilepsy and schizophrenia.

Solving these puzzles means finding answers to questions such as how neurons know where to go, which molecules guide them and why different genes are expressed in various parts of the brain.

Some of the answers will come from a global gene profile of the developing brain. They have made substantial strides using a technique called serial analysis of gene expression (SAGE) and are currently the only Australian team to master the technique. The 40,000 gene transcripts so far identified from developing cortical tissue will help identify the genes controlling brain development.

They have also successfully traced cell genealogies by transplanting embryonic cells and are credited with being first to show that columns of cortical neurons are descended from a single founder cell.

>> **www.hfi.unimelb.edu.au**
and turn to page 176 for directory details

AUSTRALIAN SCIENCE

> **ABOVE:** The Institute's modern facilities in Perth, WA.

MULTI-DISCIPLINARY APPROACH AIDS UNDERSTANDING OF CHILDHOOD DISEASE

THE TVW TELETHON INSTITUTE FOR CHILD HEALTH RESEARCH focuses on the early origins of childhood disease to ensure the best chance of prevention.

The main areas of research are Aboriginal health, asthma and allergies, birth defects, disabilities and death, infectious diseases, leukaemia and cancer, mental health and perinatal epidemiology.

Finding the causal pathways involved in these diseases is significantly increased when the Institute's scientists, with expertise in many disciplines, apply their skills and powerful new methodologies to the same medical questions. This multi-disciplinary approach is a major strength and acknowledges the enormous complexity of the social, physical and emotional environment and individual genetic variability.

Identifying the community burden of childhood disease utilises the Institute's unique maternal and child health research database which contains extensive records of births and deaths, special disease registers, hospitalisation records and census-based information.

The Institute works closely with government departments, communities and hospitals to ensure that only tried and tested policies, programs and clinical practice are implemented.

>> **www.ichr.uwa.edu.au**
and turn to page 183 for directory details

Under the leadership of University of Western Australia epidemiologist Professor Fiona Stanley, scientists with the TVW Telethon Institute for Child Health Research in Perth are tracking the causes of health problems which are major burdens to the community: asthma, adolescent suicide, birth defects, cerebral palsy, cancers and Aboriginal health. They have successfully brought their findings to public attention. For instance, their finding that inexpensive supplements of folate, one of the B vitamins, can prevent the birth of children with spina bifida, has dramatically reduced the incidence of this debilitating condition worldwide.

Not all Australian research is aimed at getting us well. A host of organisations are working at keeping us well. Scientists with the Prince of Wales Medical Research Institute not only study spinal injury, they also advise public interest groups. The NSW Motor Accident Authority and officials of Australia's various rugby codes have worked with the group in the area of prevention of spinal injury.

The POWMRI falls research program seeks to determine who is at risk of falling and injuring themselves and to prevent falls. The team has developed simple non-invasive tests which predict falls with 70 per cent accuracy; when combined with a measure of bone density, they predict fractures with 90 per cent accuracy.

Diabetic and non-diabetic people are benefiting from work conducted by human nutritionists at the University of Sydney. Associate Professor Jennie Brand-Miller and her colleagues have devised the

HEALTH AND MEDICAL

> **ABOVE:** James Lance GlaxoSmithKline Medicines Research Unit, Prince of Wales Hospital, Sydney.
> **LEFT:** Headquarters, GlaxoSmithKline, Boronia, Victoria.

PHARMACEUTICALS COMPANY'S R&D COMMITMENT PAYS OFF

GLAXOSMITHKLINE (GSK) Australia is one of the country's leading research-based pharmaceutical companies. It is committed to supporting innovative research and development for the wellbeing and economic benefit of all Australians—enabling them do more, feel better and live longer.

The company's prescription pharmaceuticals, vaccines and consumer healthcare products currently help treat and prevent disease in more than 16 million Australians.

GSK Australia is determined to remain at the forefront of the rapid progress in science and technology that will transform medical practice over the next 20 years. The company is committed to undertaking all phases of research here in Australia—from basic molecular research to final stage clinical trials.

In this respect, GSK Australia has proven its ability to take a molecule from the laboratory in Australia all the way to the marketplace on a global scale. The company has already placed Australia on the global R&D map with success stories such as Relenza™, the world's breakthrough influenza treatment, and Kapanol™, a sustained-release morphine product for the management of moderate to severe pain.

GSK Australia has helped create a world-class research base via the funding of hundreds of Australian scientists and clinical research staff. As a leading investor in pharmaceutical research and development (R&D), investment of more than $20 million per year bears testimony to the commitment of the company to the advancement of Australian research excellence.

GSK Australia is ranked number 14 on Australia's 'R&D and Intellectual Property Scoreboard', and is one of the top 20 companies contributing to total research activity in Australia. This translates into funding for more than 20 R&D discovery projects and 86 clinical trials around the country.

The company is a leader in four major therapeutic areas—anti-infectives, central nervous system, respiratory and gastro-intestinal—and the increasingly important area of preventive health through vaccines. It also has a broad consumer healthcare portfolio with over-the-counter medicines, oral care products and nutritional drinks.

In the coming years, GSK Australia will continue developing new products for unmet medical needs by working in partnership with Australian scientists and researchers.

™ is a registered trademark of the GlaxoSmithKline group of companies.

>> **www.gsk.com.au**
and turn to page 175 for directory details

AUSTRALIAN SCIENCE

ANOTHER KIND OF GOLD—THE RICHES OF AUSTRALIA'S BIODIVERSITY

An innovative joint venture partnership between ASTRAZENECA and Griffith University in Queensland has seen thousands of plants and marine organisms screened for new biologically active compounds that may be tomorrow's medical cures.

Biodiversity is at its greatest in the earth's tropical and subtropical zones. The rainforests of Queensland alone contain an estimated 9,000 plant species - 75 per cent of which are found only in Australia.

AstraZeneca R&D Griffith University (AZGU) was established in 1993 to screen the diverse flora of these rainforests and the sponges of the Great Barrier Reef. The ultimate goal is to isolate active molecules that may form the basis of new pharmaceuticals.

Since the project began, more than 24,000 samples have been collected in minute quantities and 460 compounds of interest have been identified and isolated for further testing. The screening process has utilised 74 biological assays and compounds from at least 13 of these screens have warranted closer examination and evaluation for potential pharmacological effects.

They may hold the secret to cures and treatments for a range of diseases and illnesses including heart disease, asthma, pain control, stomach disorders, migraine and cancer.

AstraZeneca recently extended their $28 million investment in this research and development collaboration by a further $37 million. This commitment to Australian research means AZGU's multi-disciplinary team of scientists can conduct research at world best-practice levels, freed from the time-consuming need to hunt for government grants and funds from private sources. It also means they have access to the high quality equipment required for this task and can provide experience and training beyond the financial capabilities of most Australian universities.

The scientific disciplines being nurtured at AZGU include advanced specialities in chemistry and biology and, on the technological side, robotics and secure data communication.

While it takes many years for a drug to pass all the necessary trials before it is available to treat patients, the signs are extremely hopeful.

>> www.AstraZeneca.com.au
and turn to page 170 for directory details

HEALTH AND MEDICAL

new 'GI' symbol—a 'G' in the centre of two concentric circles—poised to appear on food packages across Australia and North America. The symbol will alert consumers to the glycaemic index of carbohydrate-containing foods. The index ranks foods according to their immediate effect on blood sugar levels.

And thanks to the University of Sydney's Professor Colin Sullivan and his associates, people in nearly 50 countries are finally getting a good night's sleep. The listed company ResMed has commercialised their snore-busting invention—the nasal continuous positive airway pressure device (CPAP). The CPAP is the first successful non-invasive treatment of obstructive sleep apnoea, an annoying and potentially dangerous sleep disorder.

Experts across CSIRO divisions are collaborating on a telehealth initiative. This exciting program is developing communications technology to deliver health services to remote areas.

Ultimately, consultations with patients, doctor support, and other health-care services could be delivered to localities that are far from centres of medical excellence.

The Hospital Without Walls project exemplifies this concept. It aims to devise sensing and communications technology which will allow elderly and chronically ill patients to be cared for in their own homes.

When preventive programs such as these are combined with advances in the understanding of diseases, improved diagnostic techniques, and better therapies, the end product is good health.

It is a gift from Australian science, and an outcome that would please Peter Doherty.

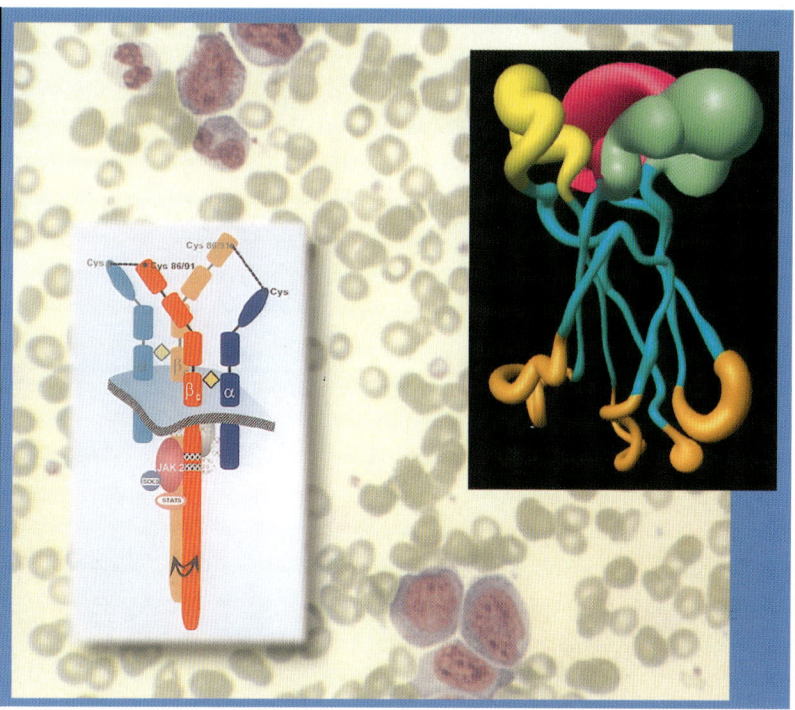

> **ABOVE:** A smear of bone marrow cells including Chronic Myeloid Leukaemic (CML) cells with an insert featuring a diagram of the human receptor for Granulocyte-Macrophage Colony Stimulating Factor (GM-CSF). The drug E21R blocks the binding of GM-CSF to this receptor inhibiting the growth of CML cells.
> **INSET:** A ribbon diagram of the X-ray crystallographic structure of the complex between the drug E21R and part of the Granulocyte-Macrophage Colony Stimulating Factor receptor. E21R inhibits the growth of a number of blood cell cancers.

NEW ANTI-LEUKAEMIA DRUG TRIALS

The success of the HANSON CENTRE FOR CANCER RESEARCH is exemplified by the development of the anti-leukaemic drug E21R. This specific antagonist of GM-CSF was discovered by Professor Angel Lopez, out-licensed to Adelaide biotechnology company BresaGen and has recently completed phase I clinical trials in Royal Adelaide Hospital. Phase II studies have started in chronic myelomonocytic leukaemia (CMML) and there are preliminary data with juvenile myelomonocytic leukaemia (JMML) in France suggesting efficacy.

The Hanson Centre for Cancer Research is dynamic, dedicated to excellence and recognised as one of Australia's premier medical research Institutes. Its 180 scientists and clinicians interact to bring discoveries into clinical practice with an annual research income of more than $10 million.

The Hanson Centre's research program ranges from basic cell and molecular biology to the development and application of new therapies. E21R demonstrates the potential for 'bench to bedside' development—an aspiration driving the whole of the Hanson Centre in cancer, biomedical and clinical research.

>> **www.imvs.sa.gov.au/hanson**
and turn to page 176 for directory details

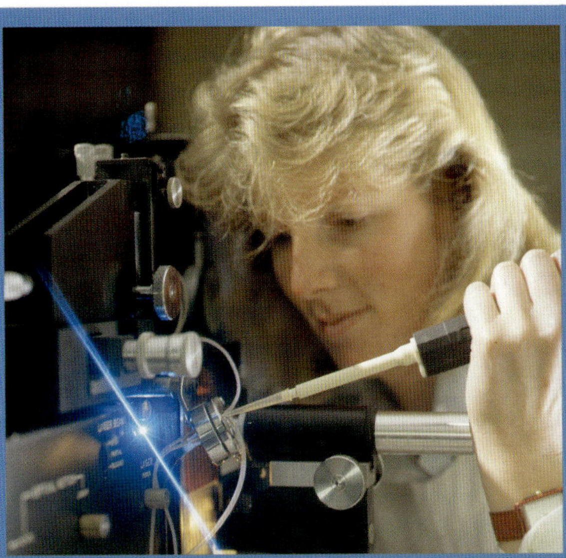

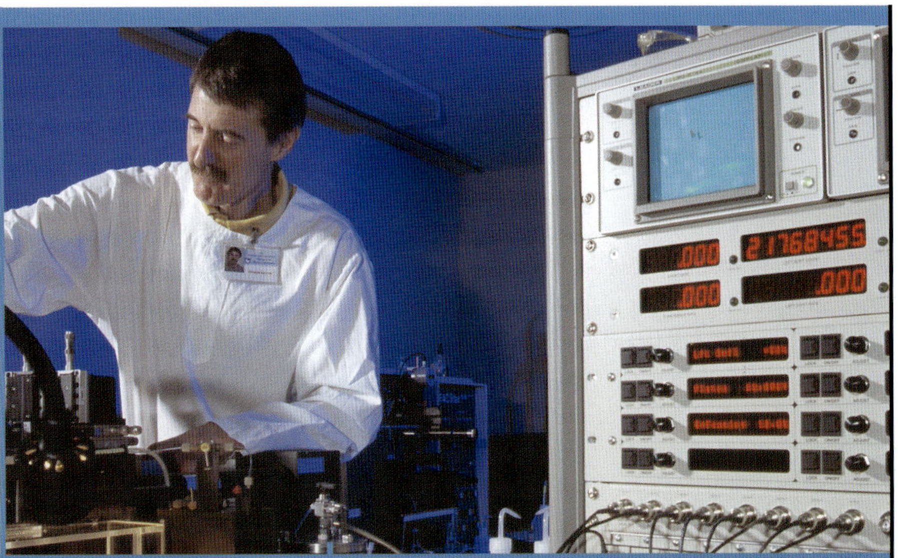

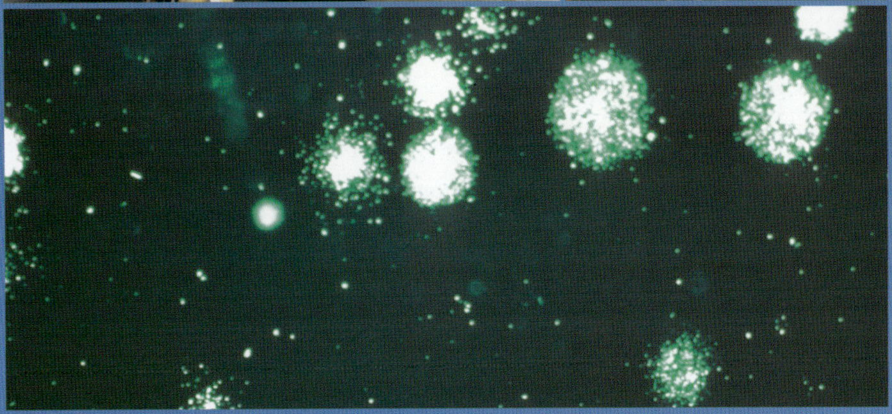

AUSTRALIAN SCIENCE

> **ABOVE:** Colonies of white blood cells grown from bone marrow cells—the basic techinque used to discover and purify the cytokines controlling blood cell formation.
> **TOP LEFT:** Kaye Wycherley injecting a cell sample into a cell sorter.
> **TOP RIGHT:** Dr Frank Battye in the FACS laboratory working on the MoFlo cell sorter.
> **RIGHT:** Professor Donald Metcalf, Carden Fellow, at work in his laboratory

WORLD LEADERS IN BIOMEDICAL RESEARCH

THE WALTER AND ELIZA HALL INSTITUTE (WEHI) is internationally renowned for research in immunology, haematology, cancer, malaria and juvenile diabetes.

Distinguished alumni include Nobel Laureate Sir Macfarlane Burnet, eminent immunologists Sir Gustav Nossal and Professor Jacques Miller, and the father of modern haematology Professor Donald Metcalf. Strong mouse genetics and new initiatives in bioinformatics, genomics and structural biology will enable WEHI to take full advantage of the Human Genome Project.

Two major areas of current research are cytokines and apoptosis. Cytokines discovered by Donald Metcalf, control white blood cell production. So far these cytokines have benefited more than 2.5 million cancer patients and revolutionised bone marrow transplantation.

Remarkably, all cells in our body can commit suicide by a process known as apoptosis. WEHI scientists discovered the gene which causes human follicular lymphoma (bcl-2) is actually a major inhibitor of apoptosis. The group is now a world leader in apoptosis research which promises new therapeutics for cancer and degenerative conditions.

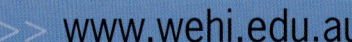

>> **www.wehi.edu.au**
and turn to page 183 for directory details

HEALTH AND MEDICAL

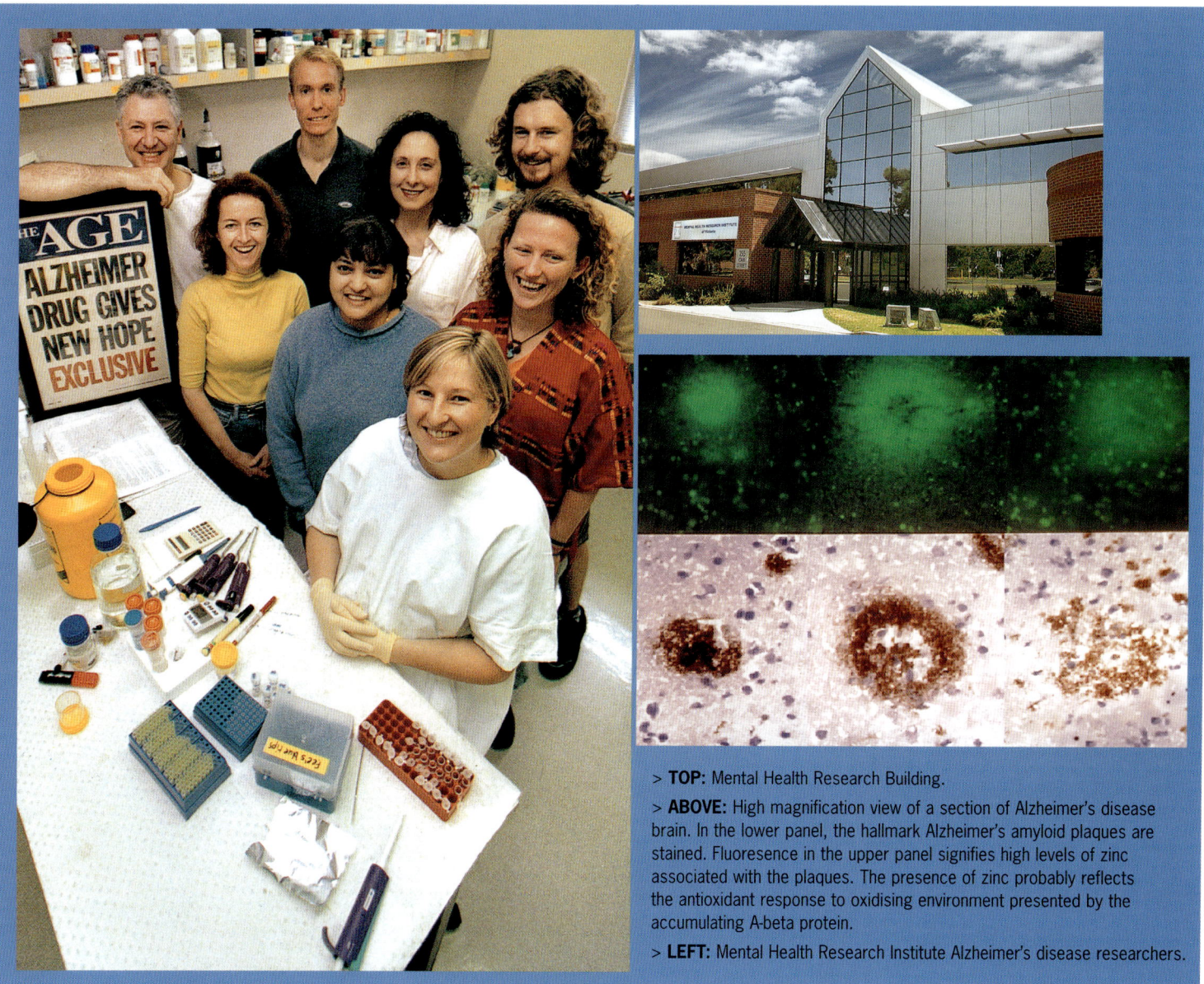

> **TOP:** Mental Health Research Building.
> **ABOVE:** High magnification view of a section of Alzheimer's disease brain. In the lower panel, the hallmark Alzheimer's amyloid plaques are stained. Fluoresence in the upper panel signifies high levels of zinc associated with the plaques. The presence of zinc probably reflects the antioxidant response to oxidising environment presented by the accumulating A-beta protein.
> **LEFT:** Mental Health Research Institute Alzheimer's disease researchers.

RESEARCHING SERIOUS MENTAL ILLNESSES

THE MENTAL HEALTH RESEARCH INSTITUTE investigates the nature, origin and causes of psychiatric diseases to improve the understanding, diagnosis and treatment of mental illness.

The main foci are serious mental illnesses such as Alzheimer's disease, schizophrenia and depressive illnesses and associated issues such as the relationship between mental illness and suicide and between cannabis and psychosis.

The Institute's Alzheimer's disease research group was the first to isolate the protein A-beta which is now recognised as the primary cause of neurotoxicity in Alzheimer's disease. Researchers are also pursuing the molecular, genetic and environmental processes implicated in neurodegeneration while collaborative research has led to successful animal trials of a drug treatment for Alzheimer's disease, with human trials now underway. The discovery of changes in the molecular architecture of the brain, in schizophrenia, are also important for the development of new anti-psychotic drugs.

The Institute has also conducted Australia's most comprehensive audit of suicides by people with psychiatric histories.

>> **www.mhri.edu.au**
and turn to page 181 for directory details

AUSTRALIAN SCIENCE

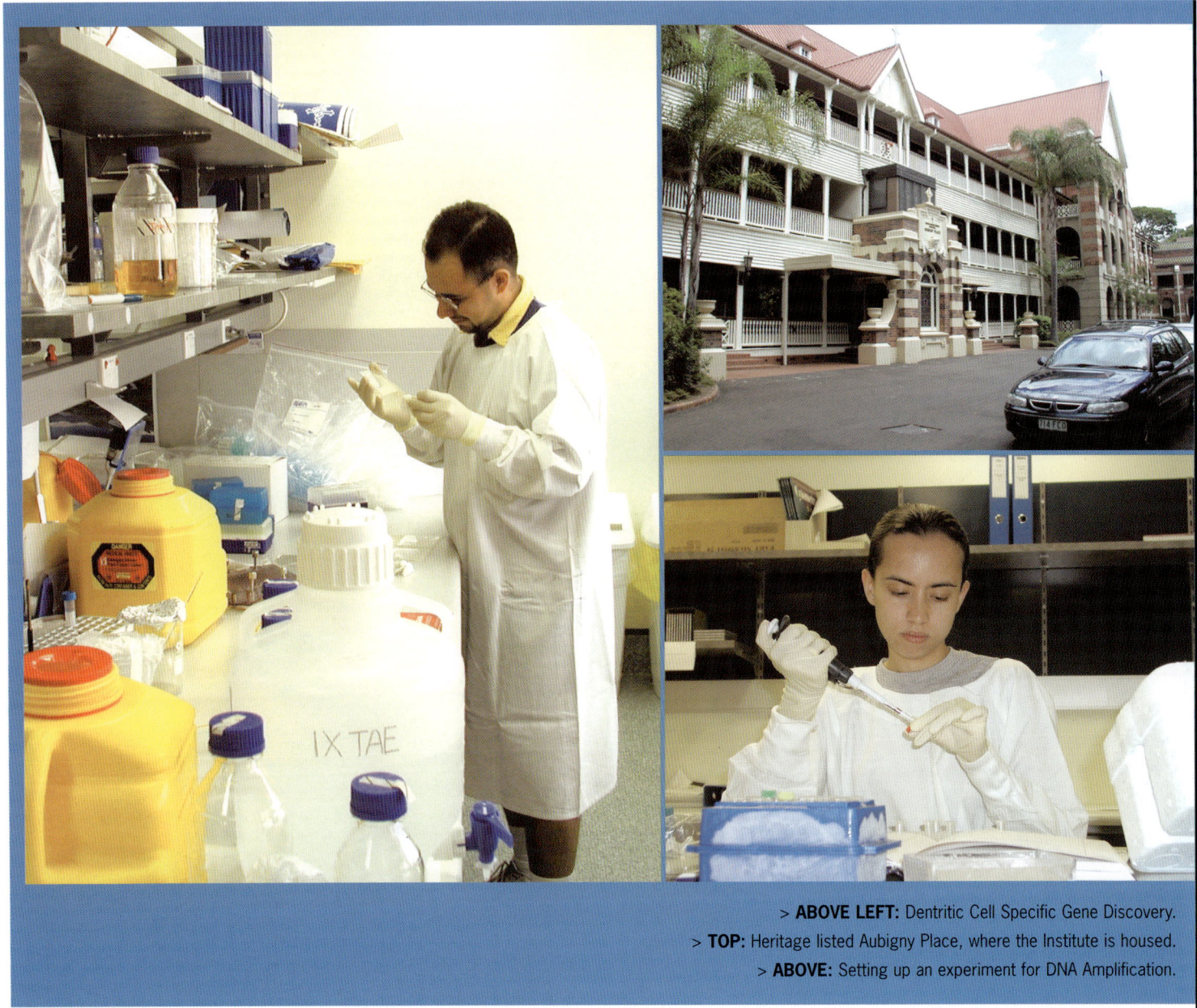

> **ABOVE LEFT:** Dentritic Cell Specific Gene Discovery.
> **TOP:** Heritage listed Aubigny Place, where the Institute is housed.
> **ABOVE:** Setting up an experiment for DNA Amplification.

COUNTING DENDRITIC CELLS IN HUMAN BLOOD

THE MATER MEDICAL RESEARCH INSTITUTE was established to be Queensland's second independent medical research institute and was developed in close collaboration with the Queensland Institute of Medical Research and The University of Queensland. Indeed the MMRI is an affiliate of The University of Queensland.

The Institute scientists are applying the advances of the cell biology revolution to therapeutic procedures designed to maximise the body's natural defenses against cancer. They have already produced the world's first test to accurately count dendritic cell's in human blood. This will allow the Institute's hematologists to predict the best time for harvesting a patient's dendritic cells for use in an effective cancer vaccine. The Institute's scientists led by the Director, Professor Derek Hart, a world authority on dendritic cells, are now developing a technology to purify these cells as a platform for cancer immunotherapy. Clinical trials have begun and it is expected that patients will benefit from this program later in 2001.

>> www.mmri.mater.org.au
and turn to page 177 for directory details

HEALTH AND MEDICAL

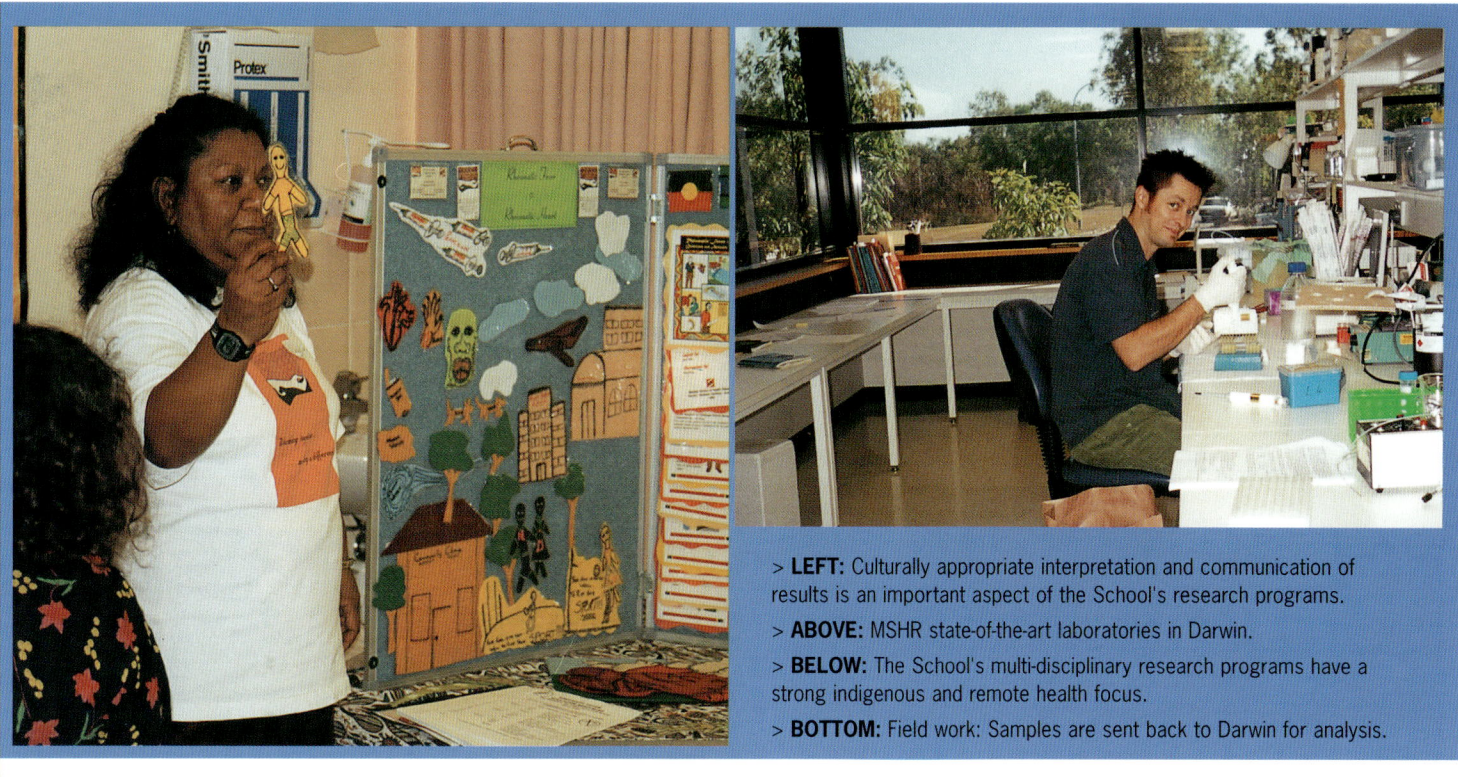

> **LEFT:** Culturally appropriate interpretation and communication of results is an important aspect of the School's research programs.
> **ABOVE:** MSHR state-of-the-art laboratories in Darwin.
> **BELOW:** The School's multi-disciplinary research programs have a strong indigenous and remote health focus.
> **BOTTOM:** Field work: Samples are sent back to Darwin for analysis.

RESEARCH PROGRAMS FOCUSING ON INDIGENOUS AND REMOTE HEALTH

THE MENZIES SCHOOL OF HEALTH RESEARCH conducts research relevant to the health of the people of northern and central Australia and countries in the region. Research spans the full spectrum from molecular to clinical research and population-based studies.

The School's multi-disciplinary research programs have a strong indigenous and remote health focus in the areas of:

> Infectious and tropical diseases—acute and chronic respiratory diseases (including otitis media), scabies, streptococcal infections and sequelae (including rheumatic fever), tuberculosis, melioidosis, arboviruses and malaria.
> Chronic diseases - diabetes, renal disease, cardiovascular disease, nutritional and other antecedents of chronic diseases.
> Social determinants of health—cultural, economic, environmental and behavioural.
> Translation of research into policy and practice.
> Interpretation and communication of research.
> Health services/systems research and evaluation.

The School's research programs cover a range of settings and modalities including the laboratory, field work (clinical, population health), international health, evidence-based clinical trials, social and behavioural science, epidemiology, public health, clinical and biomedical science.

>> **www.menzies.edu.au**
and turn to page 177 for directory details

AUSTRALIAN SCIENCE

THE PHARMACEUTICAL INDUSTRY—RESEARCH HELPS IMPROVE THE NATION'S HEALTH

The APMA is the not-for-profit professional trade association of Australia's prescription pharmaceutical industry. Its member companies are engaged in the research, development, manufacture, marketing and export of prescription medicines and the ongoing improvement of medical and scientific knowledge surrounding its products.

The APMA's vision is to be recognised as a valued contributor to Australia's health, social wellbeing and economic success, and its 50 member companies are striving to realise this vision. Every year, millions of Australians are saved from hospitalisation, surgery and death as a result of using medicines discovered by the pharmaceutical industry.

Over the past 10 years, pharmaceutical research has brought more than 300 new medicines to Australian patients. In 1998 alone, the industry delivered 36 new medicines to Australians to treat or prevent 25 diseases affecting more than six million people.

These medicines have helped reduce deaths from heart disease, cancer, AIDS and many other diseases, allowing millions of people to lead longer, healthier and more productive lives. The resulting savings to other areas of the health budget are estimated at hundreds of millions of dollars. On the research and development front, the pharmaceutical sector has boosted its research and development spending sixfold, from $42 million in 1986–87 to more than $300 million in 2000. This makes it the nation's largest non-government funder of medical research.

Collaboration with other researchers is crucial to the discovery process, so the pharmaceutical industry supports over 570 ventures with 260 medical research groups around the country, providing an average 15% of their total research budgets.

In the past 10 years, employment in this industry has risen steadily to around 13,939 people, and another 1,600 positions are funded through the industry's collaborations with other medical research groups.

Economic benefit to Australia is also a natural flow-on from this investment in local research. The result is that Australia's pharmaceutical exports have surged in the past decade from $203 million 1988–89 to an estimated $1.62 billion in 1999-2000.

>> **www.apma.com.au**
and turn to page 171 for directory details

HEALTH AND MEDICAL

> **ABOVE:** Professor David Hill, Director Cancer Control Research Institute.
> **TOP LEFT:** Professor Graham Giles, Director Cancer Epidemiology Centre.
> **ABOVE LEFT:** Professor Robert Burton, Director Anti-Cancer Council of Victoria.

LEADING IN CANCER RESEARCH, PREVENTION AND EDUCATION PROGRAMS

THE ANTI-CANCER COUNCIL of Victoria is an independent, volunteer-based charity whose mission is to lead, coordinate, implement and evaluate action to minimise the human cost of cancer to all Victorians.

In 1999, the Australian National University announced the Anti-Cancer Council of Victoria was the most successful funder of basic biomedical research in Australia during 1993 and 1994, based on citations per publication. Today, the Council's cancer research, prevention and education programs are recognised as some of the best in the world.

In 2000, the Council funded an $8.35 million research program which included biological and clinical research in hospitals, universities and research institutes as well as behavioural, epidemiological and clinical research in its own Cancer Control Research Institute (CCRI). These research funds were raised through events such as Daffodil Day, Relay for Life and Australia's Biggest Morning Tea; and from bequests, donations, trusts and other sources.

The Council's Centre for Behavioural Research in Cancer (CBRC) and the Cancer Epidemiology Centre (CEC) are the CCRI's own core research programs.

Professor David Hill, one of Australia's leading behavioural scientists, is director of the CCRI and also directs the CBRC, which investigates all aspects of the behavioural science of cancer control. This program provides the research base which underpins the Council's control strategies for lung, skin, breast, cervical and prostate cancers and the needs of cancer patients.

The CEC, under the direction of leading epidemiologist Professor Graham Giles, is currently undertaking a large longitudinal study into possible links between diet and cancer. This project, dubbed the 'Melbourne collaborative study', has been measuring dietary and other lifestyle factors in 41,500 Victorian men and women since 1990. It's hoped that the study, which includes comparative data on a variety of diets, including Mediterranean and traditional Australian, will identify possible dietary causes of cancer by recording data well before cancer is diagnosed in study participants.

>> www.accv.org.au
and turn to page 170 for directory details

AUSTRALIAN SCIENCE

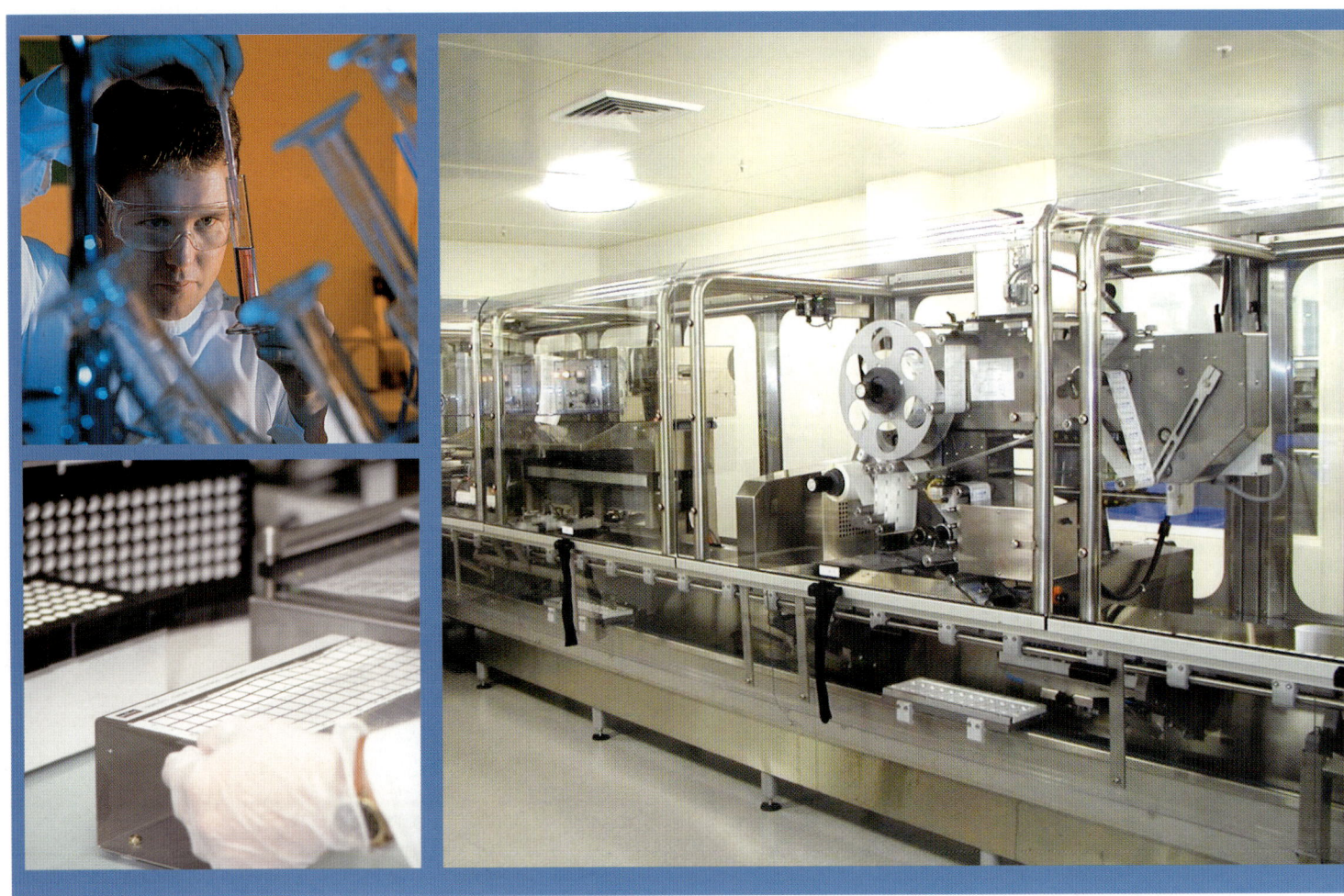

SAFE PACKAGING FOR PHARMACEUTICAL PRODUCTS

ASTRAZENECA in Australia was one of the first companies in the world to realise the potential of a new industrial technology for the production of sterile packaging of pharmaceutical products.

In 1986 AstraZeneca—known at that time as Astra Pharmaceuticals—achieved a world first by registering and commercialising the so-called 'blow-fill-seal' technology to produce sterile plastic products for use in parenteral (intravenous) feeding. This was achieved after an extensive validation and testing program that proved beyond doubt the process was capable of meeting strict requirements for sterile parenteral products.

Blow-fill-seal is an extremely efficient, high quality system, which makes a sterile container, fills it with a liquid product and seals it—all in the one operation. Because both the container and the product require no human handling, virtually all chances of contamination are eliminated. The finished container is called an ampoule.

AstraZeneca is a world leader in the move from traditional glass vials to modern plastic-based products. Not only is plastic more environmentally friendly, it eliminates the risk of glass shards contaminating sterile product or cutting the health professional's hand.

The manufacturing process means products can be tailored and packaged in varied sizes which reduces the risk of a patient receiving an incorrect dose. In addition, the containers do not require special cleaning, sterilising, storage or handling.

Blow-fill-seal technology currently produces containers for products ranging from anaesthetics and asthma inhalant medications to sterile water, sodium chloride and potassium chloride.

AstraZeneca in Australia has also been a leader in the development of other plastic parenteral products and processes including a plastic dental cartridge and a pre-filled plastic syringe with a unique method of opening, which allows easy connection directly to a syringe. Both these products were developed in Australia by AstraZeneca and are now used around the world.

>> www.AstraZeneca.com.au
and turn to page 170 for directory details

HEALTH AND MEDICAL

> **ABOVE:** Professor Chris Goodnow, Director of the JCSMR Medical Genome Centre.
> **INSET:** A new strain, generated at the Medical Genome Centre, of fat mouse (right) that mimics severe human obesity is shown next to a normal mouse.

FAT MICE HELP IN UNDERSTANDING OBESITY

THE JOHN CURTIN SCHOOL OF MEDICAL RESEARCH—Australia's national medical research school—is part of the Australian National University's Institute of Advanced Studies. The school has a 50 year history of research achievement unparalleled by any other Australian biomedical research institution.

In line with its commitment to carrying out fundamental research in the sciences with practical medical application, the school recently established the Medical Genome Centre to promote research into the function of genes that underpin human health.

Many of the health problems we face today—such as cancer, autoimmune diseases, allergy, cardiovascular disease and osteoporosis—stem from a discordance between genetics and environment.

In the last year, the Medical Genome Centre successfully screened the first of a series of libraries of laboratory mice in which parts of their genetic code have been changed by random chemical mutagenesis.

This library has yielded many new mutant mouse models which will provide greater understanding of the genes controlling obesity, cardiovascular disease, limb and spine abnormalities, eye function, coordination, seizures, dermatitis, colitis, and disorders of lymphocyte development such as autoimmunity, leukaemia and lymphoma.

This 'forward genetics' resource is being used by collaborating research groups to identify and study genes that are important in susceptibility, resistance or prevention of these disease processes.

They are using transgenesis and DNA chip technology to search for subtle changes in the behaviour of cells and tissues. In addition, state-of-the-art facilities for transgenic mouse production, sperm freezing, and database tracking of genetic and phenotypic data have been developed to support this resource and provide a service to researchers.

Other successful research underway at the school includes the anti-cancer drug 'PI-88' which is now in clinical trials, the HIV 'Co-X-Gene' vaccine technology which is about to enter large scale clinical trials, promising research into drugs which act by blocking viral ion channels, and an early detection test for cancer.

>> **www.jcsmr.anu.edu.au**
and turn to page 176 for directory details

AUSTRALIAN SCIENCE

> **ABOVE:** Chromatography columns are used to purify albumin in CSL's plasma products facility.
> **ABOVE LEFT:** Computerised systems monitor and control manufacturing processes in a CSL facility for the manufacture of novel biopharmaceuticals.
> **LEFT:** Clotting factors processing in CSL's plasma products facility at Broadmeadows in Melbourne.

ANTI-TUMOUR VACCINE ON THE WAY

Stimulating the human immune system to recognise and kill cancer cells is an attractive approach to reducing tumours, increasing quality of life and potentially curing patients with various forms of cancer.

CSL is currently bringing together a number of discoveries in immunology, DNA technology and cancer research to develop a vaccine which may achieve this goal.

The development of such a vaccine has been made possible by the discovery of tumour-specific antigens (substances which stimulate the production of antibodies) that are specifically expressed on cancer cells. These antigens provide a target that can be readily recognised, and attacked, by the human immune system.

One of these antigens, called NY-ESO-1, was first cloned in New York by The Ludwig Institute for Cancer Research. Because NY-ESO-1 is expressed in a wide variety of cancers including breast, lung, ovary, prostate, bladder, thyroid and melanoma, it will form the basis of the new CSL cancer vaccine.

This vaccine will also contain an immunological adjuvant (a substance which enhances the body's immune response when combined with an antigen) called Iscom®. Iscom is made from natural glycosides called saponins which are derived from the bark of the Quillaja saponaria (soapbark tree) found in arid areas of Chile. The Iscom® adjuvant has already been shown to produce effective immune responses in humans and even induce cytotoxic T cell responses to various viruses.

CSL's new vaccine is now undergoing clinical trials. Once they have achieved proof of principle that a combined NY-ESO-1 antigen and Iscom® adjuvant vaccine can generate an effective immune response in early-stage cancer patients, the next stage is to develop and test a therapeutic tumour vaccine.

It's hoped that the resulting vaccine will stimulate and enhance the body's immune system sufficiently for it to identify, attack and kill tumour cells in humans.

>> www.csl.com.au
and turn to page 174 for directory details

HEALTH AND MEDICAL

> **ABOVE:** Merck and Co., Inc. is developing a vaccine that could save hundreds of thousands of lives every year.
> **LEFT:** Merck Sharp and Dohme Australia—Investing in Australia's future.

MERCK SUPPORTS DEVELOPMENT OF CERVICAL CANCER VACCINE

MERCK & CO., INC* is playing a major role in developing a vaccine for the human papilloma virus (HPV), believed responsible for the development of more than 93 per cent of cervical cancer cases. After funding early work conducted by CSL Ltd and the University of Queensland, Merck & Co is continuing trials for testing the vaccine, which has the potential to prevent hundreds of thousands of deaths from cervical cancer worldwide.

HPV causes anogenital warts and is transmitted sexually. Endocervical wart infections caused by types 16 or 18 HPV have been implicated as the leading cause of cervical cancer. The discovery of the HPV vaccine began in 1989 when a research group in CSL Ltd identified HPV as a potential vaccine target. The group joined forces with Professor Ian Frazer, now director of the University of Queensland's Centre for Immunology and Cancer Research (CIRC).

Frazer's breakthrough came when he developed novel antigens called 'virus-like particles' that could be used as the basis for an HPV vaccine. CSL patented the vaccine in 1991 and was approached by a number of multinational pharmaceutical companies interested in licensing the vaccine, including Merck & Co. CSL had worked with Merck & Co. on other vaccine projects and was comfortable with the company's high ethical standards and strong capabilities in bringing products to market.

In 1995, Merck & Co. acquired exclusive rights to the virus-like particles technology and began a vaccine development program. By 1997, vaccine preparations suitable for clinical trial were available.

Merck & Co. has taken the vaccine through Phases I and II of clinical trials globally, first establishing the safety, biochemical and psychological effects of the vaccine on healthy volunteers and then focusing on efficacy and dosage rates with small patient groups. During Phase III trials, 25,000–30,000 people around the world will be tested and monitored over 2–4 years.

Merck & Co's Australian subsidiary will manage the Australian site for the final clinical trials. If the vaccine proves successful in the Phase III studies, regulatory approval for use in all patients could be sought by 2004–5.

* Merck & Co., Inc, Whitehouse Station, NJ, USA (Merck Sharp & Dohme)

>> **www.msda.com.au**
and turn to page 178 for directory details

AUSTRALIAN SCIENCE

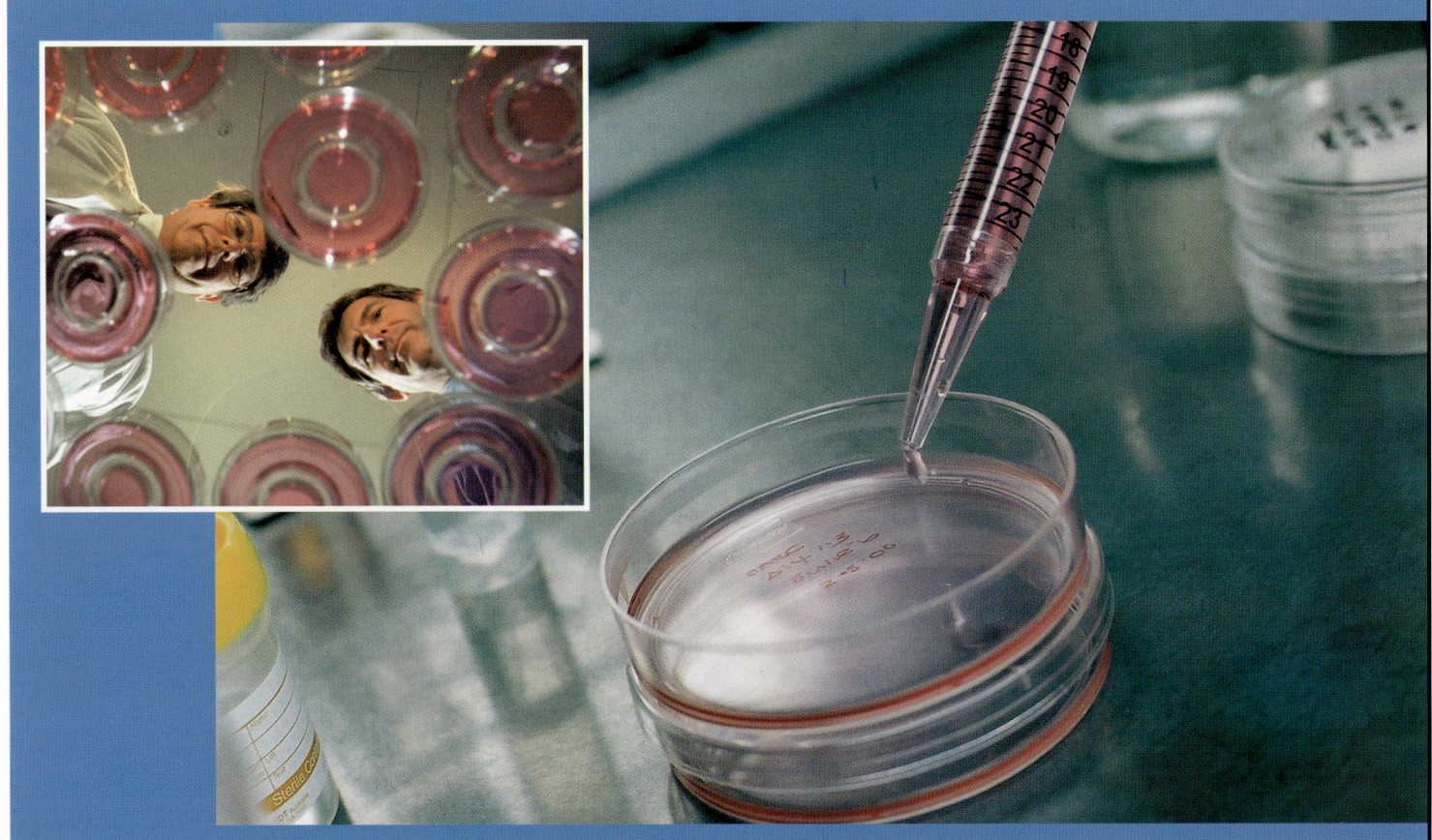

> **INSET:** Professor Alan Trounson and Dr Martin Pera from MONASH UNIVERSITY'S INSTITUTE OF REPRODUCTION AND DEVELOPMENT head a team of scientists who have grown human nerve cells from embryonic stem cells. The breakthough could lead to new treatments for degenerative diseases and offers a new means of studying human embryonic development.

CLONED NERVE CELLS COULD BECOME REPLACEMENT NERVES AND ORGANS

When scientists from MONASH UNIVERSITY'S INSTITUTE OF REPRODUCTION AND DEVELOPMENT grew human nerve cells from embryonic stem cells they made a breakthrough that could pave the way for new treatments for degenerative diseases such as Alzheimer's, Parkinson's disease and stroke.

The team's success marks the first time that cells derived from an early human embryo have been successfully turned into nerve cells in the laboratory. Those therapeutically cloned nerve cells could eventually become replacement nerves and organs which would help overcome a devastating range of illnesses.

The scientific team, led by Professor Alan Trounson, believes that human embryonic stem cells represent, in principle, an indefinitely renewable source of any type of human cell. Embryonic stem cells have the ability to turn into any type of adult tissue cell such as nerve, blood or heart cells, and can be grown in the laboratory.

They also have major applications in research and medicine offering a new means of studying human embryonic development and its disorders such as birth defects and childhood cancers, and provide a new resource for the discovery of molecules which might help in regenerating diseased or damaged tissues for organ transplantation.

The Monash team proved its theory with a mouse model. Scientists established cloned mouse stem cell lines which have the potential to grow into any type of mouse cell. The DNA was removed from an unfertilised mouse egg and replaced with the DNA of another developed cell from a 'target mouse'. This insertion of a nucleus from a developed or differentiated cell 'fertilises' the egg by introducing two complete sets of chromosomes. The next stage in the development of this therapy is to grow the cells using a patient's own DNA so they are not rejected by the immune system when introduced into the body.

>> www.monash.edu.au
and turn to page 178 for directory details

HEALTH AND MEDICAL

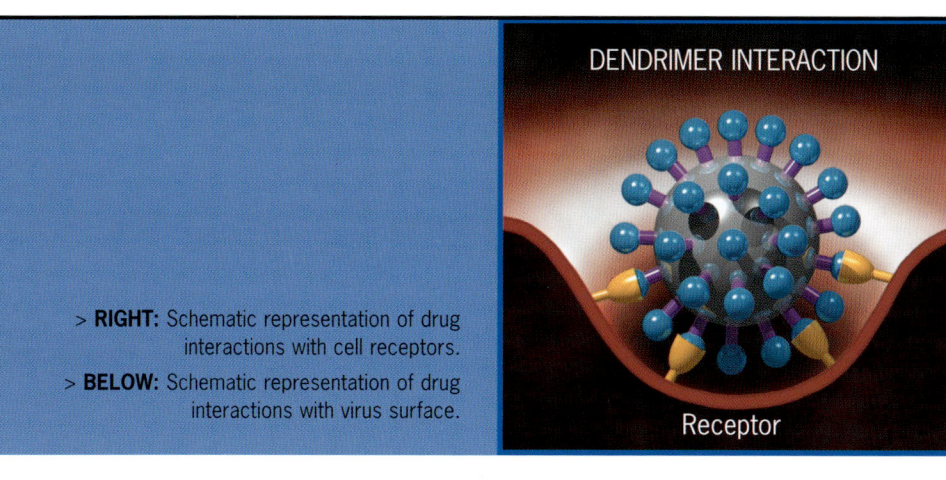
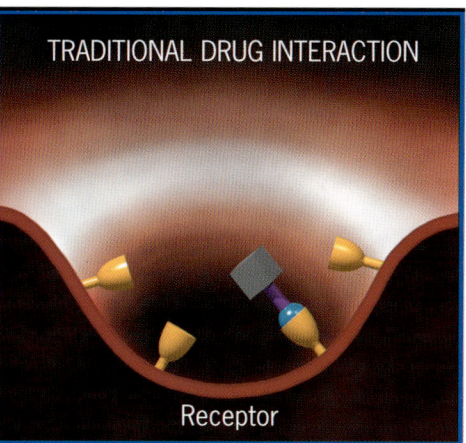

> **RIGHT:** Schematic representation of drug interactions with cell receptors.
> **BELOW:** Schematic representation of drug interactions with virus surface.

USING DENDRIMERS TO TREAT DISEASE

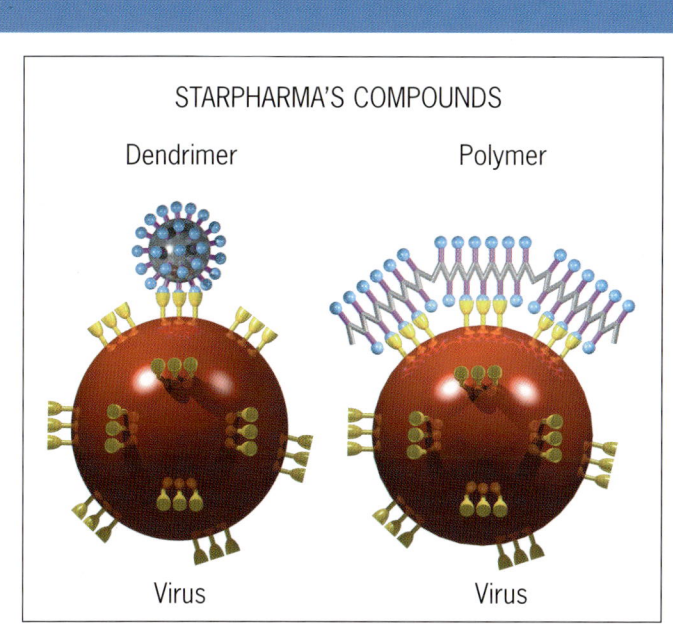

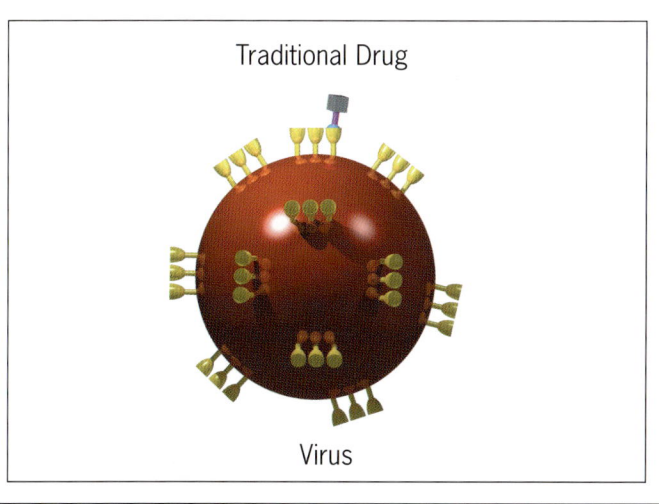

STARPHARMA POOLED DEVELOPMENT took its name from the company's platform technology of multibranched, 'star shaped' chemical entities known as dendrimers. These structures are much larger than traditional drugs and interact with biological receptors through multiple contacts (polyvalency). Different surface groups can be attached to the core scaffold to achieve activity in different disease applications. Unlike polymers, the size and composition of the dendrimers is precisely known and does not vary from batch to batch. In preclinical testing, the Starpharma dendrimers have shown remarkable activity against viruses, solid tumour metastases and toxins.

Starpharma's lead project is the development of a vaginal microbicide, effective against the incurable sexually transmitted viral diseases, namely HIV, Herpes Simplex Virus (HSV 1&2), Human Papilloma Virus (HPV) and Hepatitis B (HBV). The broad spectrum and novel antiviral activity of the dendrimers particularly lends itself to the development of such a product. The lack of woman-controlled protection represents a critical unmet reproductive health need worldwide. With the advent of AIDS and the spiralling epidemics of HPV and HSV which promote the spread of HIV, the need is acute.

Starpharma's other lead project is in the field of angiogenesis inhibition. Dendrimers are being evaluated to prevent or treat a range of diseases associated with the process of new blood vessel formation (angiogenesis). Potential disease targets in this field include cancer (inhibition of metastases and solid tumour growth), arthritis, retinitis and asthma. Starpharma is also developing a class of small chemical entities that have shown high level inhibition of angiogenesis. This class is being progressed through preclinical development along with the dendrimers.

A dendrimer with combined antiviral and anti tumour activity is about to enter woodchuck trials in the USA to assess whether there is an effect on hepatitis B and the liver cancer that is induced by this virus.

>> **www.starpharma.com**
and turn to page 181 for directory details

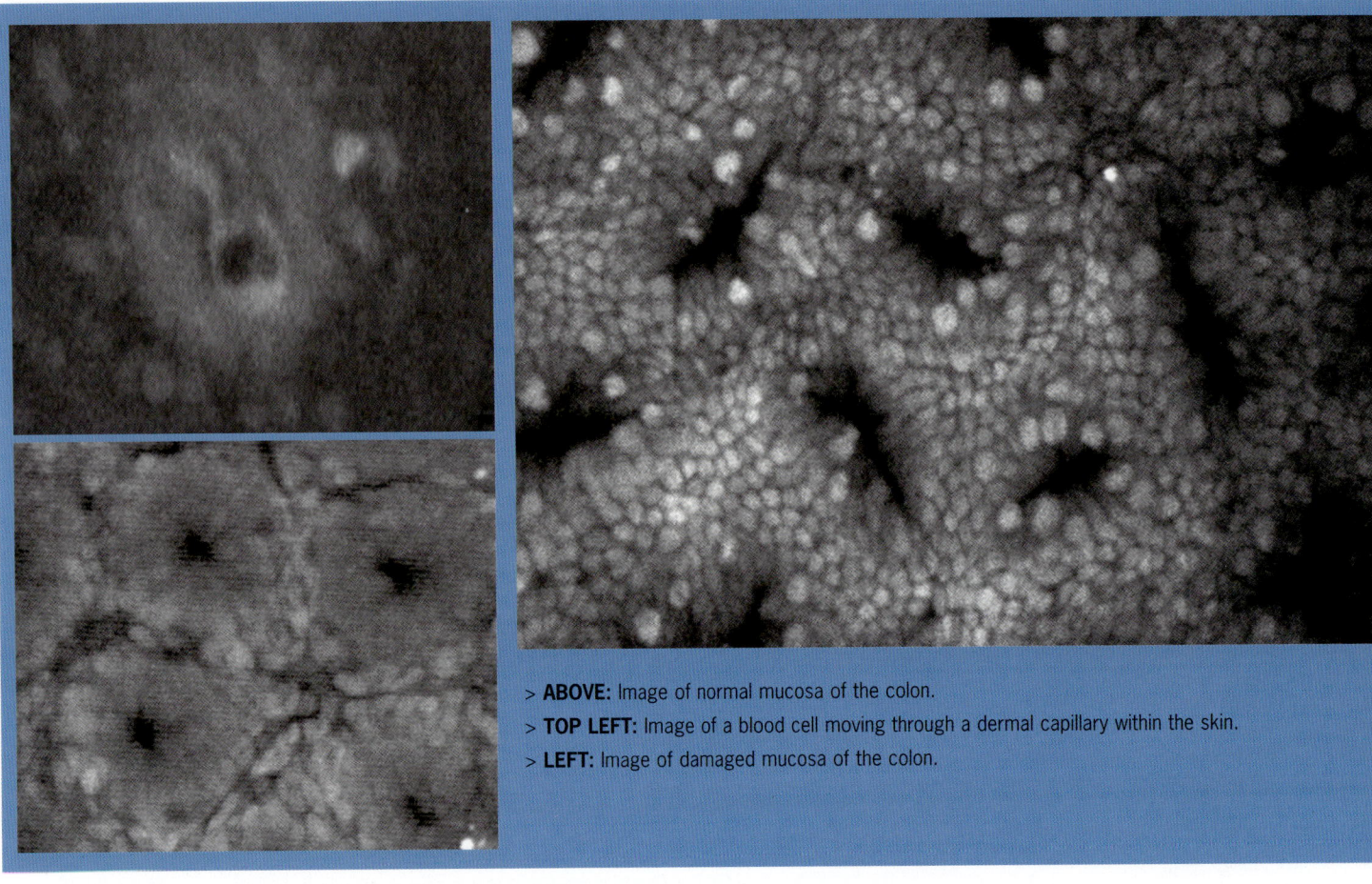

> **ABOVE:** Image of normal mucosa of the colon.
> **TOP LEFT:** Image of a blood cell moving through a dermal capillary within the skin.
> **LEFT:** Image of damaged mucosa of the colon.

QUICK, EASY AND NON-INVASIVE MELANOMA EXAMINATION

OPTISCAN Imaging Limited has invented the world's first hand-held confocal microscope for use in the early detection of skin cancer. In a major leap in cancer prevention and early detection, cells in potential cancers can now be examined non-invasively, without the need for skin biopsies.

The hand-held microscope is simply placed against the skin on any part of the body. It uses visible and/or near infrared light to non-invasively produce high magnification (x1000) images of microscopic cellular structures. Abnormal cellular changes can then be detected using the hand-held microscope.

This new device can detect abnormal cells at the very early stages of disease, when there are just a few cancer cells present. This offers an accurate observation to be made well in advance of that of the conventional methods of clinical diagnosis. The crucial factor in support of this system is that early detection of a skin cancer, such as melanoma, means a 98 per cent chance of survival compared to that of ~5% survival when detected in advanced stages.

Benchtop models of the device are being used by researchers at sites including Monash and Sydney universities, the CSIRO and the Royal Children's Hospital in Melbourne. The hand-held clinical version will soon be available to hospitals and pathologists.

Optiscan Imaging Limited is a medical research, development and manufacturing company based in Melbourne. Although the technology behind the Optiscan device has a wide range of potential applications, the company has concentrated on what it believes to be the most compelling use - medical examination.

There are significant worldwide markets for non-invasive microscopic medical examinations due to the high cost of curative medicine. Other potential uses for the Optiscan technology include monitoring of psoriasis and other inflammatory skin conditions, gastrointestinal disorders, cervical cancer screening, monitoring of cellular responses in cancer therapy and cosmetic surgery.

Optiscan Imaging is currently developing an endoscopic version of the device to be used in the examination of early stage cancers of the colon, cervix, ear, nose and throat.

>> www.optiscan.com
and turn to page 179 for directory details

HEALTH AND MEDICAL

> **ABOVE AND TOP RIGHT:** The new centre will make QIMR the largest medical research facility in the Southern Hemisphere.
> **BELOW RIGHT:** Cells lining a crypt in the colon. QIMR investigates the molecular and genetic factors that lead to colon cancer.

UNIQUE CANCER AND CLINICAL RESEARCH CENTRE FOR QIMR

THE QUEENSLAND INSTITUTE OF MEDICAL RESEARCH (QIMR) in Brisbane is taking a revolutionary step forward in Australian cancer research with the opening of the Comprehensive Cancer Research Centre.

The new centre offers the unique combination of basic cancer research alongside the development of new cancer treatments. It will further expand QIMR's world-renowned expertise in cancer research and create the largest medical research centre in the Southern Hemisphere.

The centre's gene therapy facilities, clinical trials unit and state-of-the-art laboratories will house more than 400 scientists working in close partnership with researchers and clinicians in Australia and around the world.

QIMR has a proud history of success in the discovery of cancer-causing agents and the development of approaches that may help prevent the disease.

Over the years, its scientists have played an important role in the discovery of genes responsible for cancer, and discovered that sunblock creams have the ability to prevent some types of skin cancer.

The world's first clinical trial of an immunology cancer vaccine, carried out at QIMR in 1994, used gene therapy to enhance a patient's immune response against cancer. The trial's success was the basis of ongoing development of cancer treatments that are effective, yet gentle enough not to compromise quality of life.

Currently the QIMR is trialling a new cancer treatment in which an immune response against cancer cells is grown in tissue cultures and then infused into cancer patients.

The completion of the new cancer research centre will enable QIMR to continue developing drugs and vaccines of the highest quality.

Major ongoing research programs at QIMR—including gene research, immunology and vaccine development, epidemiology, liver diseases and transplantation, tropical and infectious diseases, molecular and cell biology, and indigenous health—will guarantee a leading role in medical research for the future.

>> **www.qimr.edu.au**
and turn to page 182 for directory details

AUSTRALIAN SCIENCE

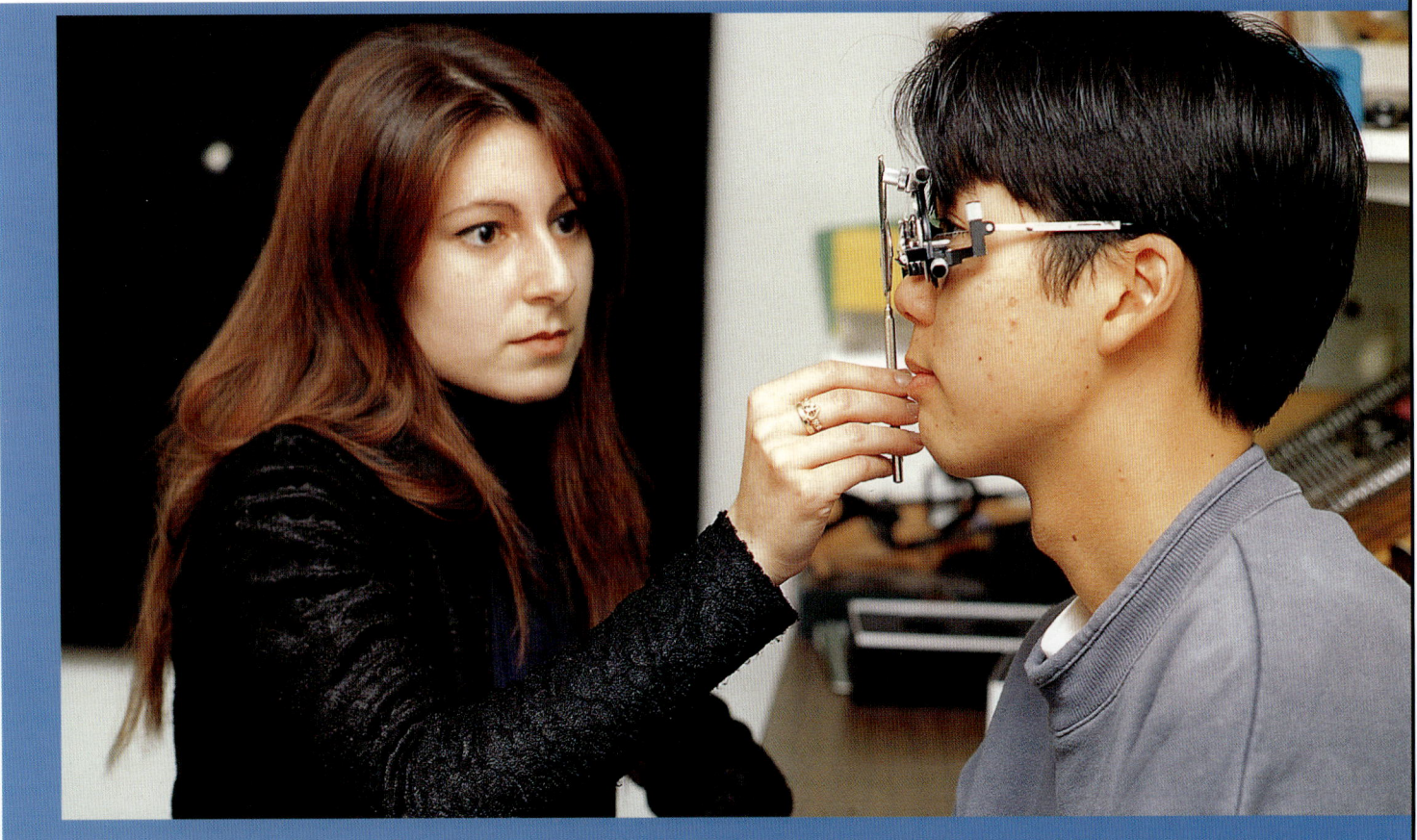

> **ABOVE:** Researchers at Sydney University—leading the world in research to combat degenerative eye disease

TRIALS OF OXYGEN TO LIMIT RETINAL DEGENERATION

THE NSW RETINAL DYSTROPHY RESEARCH CENTRE at the University of Sydney is committed to furthering our understanding of degenerative diseases of the retina.

Directed by Professor Jonathon Stone, Challis Professor and Medical Foundation Fellow at the University of Sydney, the centre's work largely concerns the diseases retinitis pigmentosa, age-related macular degeneration, retinal detachment and retinopathy of prematurity.

In collaboration with the Neuroscience Research Institute at the University of California Santa Barbara, the centre has been investigating the role hypoxia plays in retinal detachment and the potential for using oxygen to mitigate that process.

When this research was published in the *American Journal of Ophthalmology* in early 1999, it was also featured in the journal's editorial pages. This coverage evoked an immediate response from the ophthalmology community with many wishing to take part in the clinical trials.

The phase 2 clinical trials, which commenced in 2000, are the centre's first direct involvement in therapy-related studies and the first ever trial of oxygen to limit retinal degeneration.

Another research project focused on the role of perinatal hypoxia in non-familial retinitis pigmentosa commenced with Australian patients in 1998. This study was then extended to include patients at the Jules Stein Eye Institute at the University College of Los Angeles in the United States.

The success of this work has led to the further exploration of hypoxia resistance in some of the highly characterised inbred strains of mice developed at the Jackson Laboratories in Maine in the United States. The goal is to identify oxygen-related genes which may underlie the susceptibility some sufferers have to perinatal stress in retinitis pigmentosa.

The centre seeks to use its research for the benefit of those suffering retinal degeneration by developing therapy, identifying opportunities for prevention and meeting clinicians who claim successful therapy and critically assessing their claims.

>> www.usyd.edu.au
and turn to page 182 for directory details

Medical research today uses both high-tech and more traditional tools—the petri dish is still essential

Gene technology is being used to breed flower species with improvements that have commercial value—blue flowers and extended vase life are important focuses of this work

BIOTECHNOLOGY

BIOTECHNOLOGY

AUSTRALIAN BIOLOGICAL SCIENTISTS ARE SEARCHING FOR GENES AND BIOCHEMISTRIES THAT WILL PROVIDE NEW AND BENEFICIAL PRODUCTS AND PROCESSES, REPORTS LEIGH DAYTON

For 200 years, Australia paid its bills by raising sheep and cattle, growing food crops, and digging up valuable resources hidden in the worn and rugged landscape. Now, like 21st century farmers and miners, Australian scientists are tapping a new source of wealth—the machinery of life itself. Wielding the tools of biotechnology, they are poised to create useful processes and products, and generate billions of dollars in revenue, all from the ultimate natural resource.

But before scientists can wring profits from bacteria, bugs, or bits of biological activity, they have to go prospecting. Across Australia, a host of teams are investigating a diverse range of organisms, from fungi to viruses, people to plants. For example, CSIRO Entomology has joined with the company BioDiscovery to go bug hunting. CSIRO will collect insects and develop a 'library' of extracts from them; BioDiscovery will screen the extracts for biologically active, beneficial compounds.

Cerylid, in Melbourne, has been busy screening its collection of over 400,000 extracts from Australia's native flora in collaboration with state governments, indigenous communities and commercial companies. As the commercial partner for the Cooperative Research Centre (CRC) for the Discovery of Genes for Common Diseases, Cerylid is looking at exciting possibilities for drug discovery.

AUSTRALIAN SCIENCE

> **ABOVE:** APAF is at the international forefront of proteome technology development and application. This advantage presents an excellent opportunity to develop quality Australian based research, technology and knowledge at a commercial level, and to retain the associated benefits in Australia.

TEARS USED TO DIAGNOSE CANCER

THE AUSTRALIAN PROTEOME ANALYSIS FACILITY (APAF) has a long-standing collaboration with the Cooperative Research Centre for Eye Research and Technology which led to the possibility of a diagnostic test for cancers using tears. Tears from patients with a range of cancers at various stages of development are now being surveyed to determine the potential of this non-invasive technique to detect early stage cancer. While studying tears, a novel protein called lacryglobin was discovered which has a high degree of homology with mammaglobins that are sensitive and highly specific markers for breast cancers.

APAF was created under the Major National Research Facility scheme. It provides a service to Australian and international research organisations in the platform technology called proteomics which promises to be the next revolutionary biotechnology in the post-genomic era. APAF had more than 120 clients in 2000 covering all areas of biological science including medicine, agriculture, microbiology and diagnostics.

>> www.proteome.org.au
and turn to page 171 for directory details

At the University of New South Wales, scientists have turned the bacterium *Bacillus sphaericus* into a powder that fights malaria. When sprayed onto stagnant water where disease-bearing mosquitoes breed, *B. sphaericus* gobbles the mosquito larvae. And researchers at Macquarie University have 'milked' secretions from glands in the Australian bull ant which kill *Staphylococcus aureus*, or golden staph bacteria. This frequently antibiotic-resistant bacterium is widespread in hospitals, where it causes life-threatening infections.

Biotechnology can also make a dangerous creature, such as the bull ant, its own worst enemy. By developing antivenenes to some of the most venomous creatures in the world, University of Melbourne Associate Professor Struan Sutherland has saved countless lives. His hit list includes the deadly funnel-web spider and the tiger snake.

New, automated technologies make locating and exploiting such natural resources faster and cheaper. The Australian Genome Research Facility (AGRF), for example, is a non-profit, state-of-the-art facility for the collection of information about genes. The AGRF's University of Queensland division uses high-speed DNA sequencing equipment to study the genes found in viruses, bacteria, protozoa and fungi, and plants, animals and humans. The Melbourne division, located at the Walter and Eliza Hall Institute of Medical Research, specialises in genotyping and mutation detection and analysis.

Australian scientists are also pioneers in the emerging field of proteomics. In fact, the name 'proteomics' was coined by Dr Keith Williams, of Sydney-based Proteome Systems. Proteomics is the study of 'proteomes', collections of different proteins in individual cells. As the genome is the complement of genes of an organism, the proteome is the sum of a cell's proteins.

Like house builders, proteins are complex molecules which do the bidding of genes, the building's architects. And as 99 per cent of all drugs are either proteins themselves or act by binding to proteins, understanding proteins and proteomes will help medical researchers create new pharmaceuticals.

Proteomics will also help clarify the linkages between genes, diseases and individual responses have identified 'marker' proteins for forms of prostate, bowel, and various forms of childhood cancer. Drug designers can now structure compounds to target the markers, and the diseases.

APAF researchers are also looking for proteins which make the bacterial pathogen *Pseudomonas aeruginosa* more potent and resistant to antibiotics, and analysing other major disease-producing microbes, including gut-dwelling *Escherichia coli*, *Staphylococcus aureus*, and *Helicobacter pylori*, the 'bug' first linked to ulcers. Two West Australian doctors, Barry Marshall and Robin Warren, made that ground-breaking connection. Other proteome

PROTEOMICS WILL HELP CLARIFY THE LINKAGES BETWEEN GENES, DISEASES AND INDIVIDUAL RESPONSES TO THOSE DISEASES

to those diseases, all of which will lead to better treatments, and assist in the development of powerful products and processes for use in agriculture, farming and aquaculture. Projects aimed at improving wheat strains and sexing farm animals are already on the go.

Macquarie University's Australian Proteome Analysis Facility (APAF) houses what is currently the most advanced proteomics equipment available. APAF projects with the biotechnology firms Bio-Rad, Gradipore, Beckman Instruments, Australian Rapid Robotics Manufacturers and Hewlett Packard have led to the development of instruments and products which are now being sold internationally.

To date, no commercially available drugs have come out the APAF pipeline, but its investigators, working with Australian and overseas collaborators, projects, related to anti-allergy medications, and to extended-wear contact lenses which fit comfortably with human tears, are also under way.

A unique family of proteins called Suppressors of Cytokine Signalling proteins, or SOCS proteins, is the target of exciting work conducted by the CRC for Cellular Growth Factors. The CRC is a joint effort of the Walter and Eliza Hall Institute of Medical Research, the Ludwig Institute for Cancer Research, CSIRO Health Science and Nutrition, the Biomolecular Research Institute, and AMRAD Operations Pty Ltd. Patented SOCS proteins promise to boost the microbe-fighting power of existing therapies which target viruses such as the Human Immunodeficientcy Virus (HIV).

Once these compounds tumble from the laboratories, more cutting-edge technology goes to work.

AUSTRALIAN SCIENCE

PICTURE: WALTER AND ELIZA HALL INSTITUTE

> **ABOVE:** Dr Frank Battye in the FACS laboratory working, on the MoFlo cell sorter

Following the development of Relenza, the world's first flu drug, other new substances in the pipeline promised to give Australian scientists an impressive reputation for drug design. Relenza is based on pioneering protein crystallography conducted in the 1980s by the CSIRO. Relenza has been approved for use in the United Kingdom, the United States and Australia. GlaxoSmithKline have commissioned a manufacturing plant for the drug, in Boronia, Victoria.

In South Australia, BresaGen is also building a manufacturing plant, for its biopharmaceuticals. Established in 1982 by University of Adelaide scientist Professor Bob Symons, BresaGen has numerous 'biotherapeutics' in the pipeline—for myeloid leukemia, breast cancer, allergic disease and organ disease—and is developing veterinary products.

At the University of Queensland's Institute for Molecular Bioscience (IMB), researchers have isolated an extract from the mollusc inside the cone

A joint venture between the computer firm Advanced Micro Devices and the CRC for Cellular Growth Factors has produced a computer that is purpose built for designing new pharmaceuticals.

WITH ITS OUTSTANDING SCIENTISTS, FIRST-RATE LABORATORIES AND UP-TO-THE-MINUTE RESEARCH METHODOLOGIES, AUSTRALIA IS BUILDING AN IMPRESSIVE REPUTATION FOR DRUG DESIGN

Located at the CSIRO's Parkville laboratory, near Melbourne, the high-speed computer, called Caduceus, can screen millions of chemical compounds. Caduceus is looking for 'candidate' drugs. To find them, it compares the shape and chemical characteristics of selected compounds with those of target sites on proteins which are associated with diseases such as cancer and arthritis. This gives chemists a huge head start in the race to create safe, effective pharmaceuticals.

shell which inhibits pain. The AMRAD Corporation has licensed the compound, which is currently undergoing clinical trials. Researchers with the IMB have also isolated genes that cause or modify serious diseases. Future drugs spun off from these could reduce the severity of cystic fibrosis and of the most lethal skin cancer, melanoma. And IMB researchers envision a skin cream that could be used to control basal cell carcinoma, the world's most common skin cancer.

BIOTECHNOLOGY

As well, IMB has patented a gene for the development of blood vessels, a process known as 'angiogenesis'. Because cancers must grow blood vessels to thrive and spread, the gene promises to be the key to a powerful anti-angiogenesis drug. Used in conjunction with conventional cancer therapy, such a drug could slow or halt the spread of many solid tumours, such as breast or colon cancer, without harming healthy tissue.

The war against cancer has taken a surprising turn in the laboratory of Professor Chris Parish, at the Australian National University (ANU). His team has discovered that a cheap and simple sugar compound, an oligosaccharide, can stop the spread of many cancers. Australian biotechnology company Progen, in conjunction with the ANU team, has begun human trials of the anti-tumour sugar pill, PI-88. Like IMB's experimental drug, PI-88 works its magic without killing normal cells.

Another discovery that began life in a laboratory at ANU is about to begin trials with patients. This time the target is HIV. Working with a fowlpox virus (FPV) bioengineered by the CSIRO, ANU's Professor Ian Ramshaw has created a vaccine which may boost an HIV-infected patient's disease-fighting immune system. It works by tricking the body into producing two types of immune system products—cytokines and antigens. In Melbourne, Virax Holdings is manufacturing the vaccine (VIR201) for pre-clinical and clinical trials.

These trials have been supported by an A$27 million grant from the US National Institutes of Health. If successful, VIR201 would be made available, mostly in developing countries, to people

> **ABOVE:** Victoria—The Victorian Government recognises that science, technology and innovation are key drivers of future prosperity and quality of life.

GOVERNMENT BRIDGES GAP BETWEEN SCIENCE AND BUSINESS

THE VICTORIAN GOVERNMENT is committed to developing the science, technology and innovation base in the State, and to ensuring that Victoria is recognised as a leader in these fields locally and internationally.

The Government is funding a $400 million precinct at Parkville in Victoria, Bio21, which is intended to become a world-leading cluster of medical and scientific research institutes in the growing biotechnology industry. Foundation partners are The University of Melbourne, The Walter & Eliza Hall Institute of Medical Research and The Royal Melbourne Hospital.

The Victorian Government's technology commercialisation program is a $20 million program which bridges the gap between great ideas and commercialisation by providing inventors and innovators access to high quality professional business support and development services.

The Government also offers Science, Technology & Innovation (STI) infrastructure grants. The Victorian Government has provided $54 million funding for innovative infrastructure proposals. The first 15 projects, worth over $225 million, cover medicine and health, agriculture and food, materials science and telecommunications.

>> **www.innovation.vic.gov.au**
and turn to page 180 for directory details

AUSTRALIAN SCIENCE

PICTURE: FLORIGENE FLOWERS

> **ABOVE:** Molecular breeding of flowers is big business—the worldwide market for blue flowers is estimated to be worth US$5 billion annually. Scientists working in this field use genetic engineering to isolate the genes they need, then implant the gene from flowers that have a blue gene into flowers that don't.

who are newly infected with HIV—vaccines are thought to offer the best, most affordable, prospects for managing the HIV/AIDS pandemic.

The Western Australian Biomedical Research Institute (WABRI), a joint initiative of Curtin University of Technology and Murdoch University, is conducting drug identification to control and treat the disease trypanosomiasis, known as 'sleeping sickness' in Africa and 'surra' in Southeast Asia. This parasite affects agriculture production and threatens Australian agricultural and native animals. Research by WABRI, the WA Chemistry Centre and Glaxo SmithKline has identified new targets—several anti-parasite compounds are undergoing evaluation.

Serendipity and science frequently go hand in hand, especially in medical research. It was purely by chance that Associate Professor Philip Hogg, of the University of New South Wales (UNSW), discovered a compound that fights cancer and HIV/AIDS. Studying the way HIV enters immune system cells and hijacks their machinery to replicate, he realised that the same chemical reaction occurs during angiogenesis. Hogg then teamed up with UNSW chemist Dr Neil Donoghue to create a prototype drug. The pair recently joined forces with Australian pharmaceutical maker the Institute of Drug Technology. Their drug, GSAO, will soon be tested on patients.

In the future, patients, researchers, and a host of environmental experts will benefit from another Australian advance: the world's first functioning nanomachine. The tiny device is a biosensor, which can detect minuscule amounts of substances—it could detect the change in sugar content of Sydney Harbour after one sugar cube was thrown into it!

BIOTECHNOLOGY

>> ENGINEERED FOODS

HUMAN BEINGS HAVE USED BIOTECHNOLOGY TO MODIFY THEIR FOOD FOR TENS OF THOUSANDS OF YEARS. NOMADIC HERDERS RELIED ON A BACTERIUM, *LACTOBACILLUS*, TO MAKE YOGURT AND CHEESE FROM MILK, CURDLED BY THE DIGESTIVE ENZYME RENNET. ANCIENT EGYPTIANS BREWED BEER WITH THE HELP OF YEAST

Yeast itself is a biological compound, consisting of cells of ascomycetous fungi that clump together in a yellow, frothy, viscous growth. When yeast meets saccharine liquid, the result is brewing beer or rising bread.

Early farmers 'genetically' engineered their crop plants through domestication. Wheat, rice and rye, barley, oats, and corn look little like their parent strains. Our ancestors 'created' food and pack animals, using selective breeding to alter the genes of their wild progenitors.

During the past century, advances in biotechnology have dramatically sped up the process. But today's genetically modified foods, or GM foods, rely on the same basic principles: microbial activity, and changes to the genetic makeup of a plant, animal, or ingredient. The results are useful, and sometimes surprising.

Tomatoes have been tweaked, by reversing a single gene within the plant's genome, so that they survive handling better and stay riper longer. Bananas have been persuaded to produce vaccines, eliminating the jab of a needle while simultaneously providing a nutritious snack. And rennet is made in the laboratory, not in primitive vessels made from the lining of an animal's stomach.

Scientists have altered the genomes of many food crops in order to create strains with desirable traits. Key traits are tolerance of heat, salt, and drought, bigger yield, faster growth, and even enhanced resistance to insect pests or disease-producing organisms.

Genetic engineers have a range of tools at hand, although it usually requires only small changes to achieve their goals. They can switch genes 'on' or 'off', or 'turn up' or 'turn down' their activity. They can 'cut and paste' genetic material within the genome of a plant or animal. They can even move genes from one strain or species to another.

Australian scientists are among the leaders in the production of GM foods, but their wares are not always welcomed by the public. Australians are debating the pros and cons of GM foods. And that is important, because in the laboratory, biotechnology is ethically neutral.

Biotechnology holds the promise of increasing food supplies worldwide or reducing the need for herbicides, and the threat of unanticipated environmental hazards or human allergies. Its future depends on how we choose to use it. It is up to us.

AUSTRALIAN SCIENCE

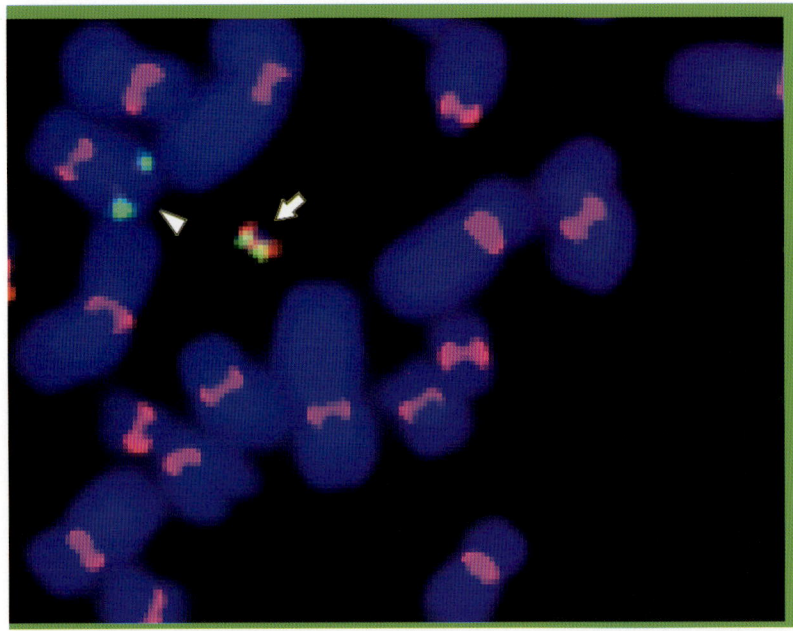

> **ABOVE:** Normal human chromosome 10 is indicated by an arrowhead and the HAC an arrow. The neocentromere-based HAC is eminently smaller in size when compared to other human chromosomes. The presence of a centromere protein (red signal) indicates that it is a fully functional mini chromosome. The green signal represents the neocentromere DNA found on both the HAC and chromosome 10.

AN INTERNATIONAL PATENT ON HUMAN ARTIFICIAL CHROMOSOMES

Professor Andy Choo's laboratory at the MURDOCH CHILDRENS RESEARCH INSTITUTE holds an international patent on human artificial chromosome (HAC) technology. This technology has enormous commercial potential and could save millions of lives through gene therapy. Gene therapy is the treatment of disease by introducing corrective copies of genes, for example introducing cancer repressor genes into cancer patients. HACs offer a promising approach to deliver genes into such patients because there is theoretically no upper limit to the size of DNA that can be incorporated. Furthermore, the use of human DNA-derived HACs, rather than viruses, as vectors should minimise the risk of adverse immunogenic responses in the patients.

One route to generate a linear mammalian artificial chromosome is by fragmenting an existing chromosome. Professor Choo's group has successfully produced a series of HACs by truncation of a chromosome 10 marker containing a human neocentromere (a fully functional centromere that arises spontaneously). The HACs made in this way demonstrate high stability in both structure and mitotic transmission over the generations. Having a minimal sized neocentromere-based HAC not only facilitates the delivery of therapeutic genes into patient cells, but also allows full sequence characterisation of the HAC to provide a better-defined tool for future gene therapy.

>> **www.murdoch.rch.unimelb.edu.au**
and turn to page 179 for directory details

It consists of a membrane, attached to a thin metal film, coated onto a piece of plastic. When it contacts target molecules, it opens or closes minute channels in the membrane.

The biosensor was created by Dr Bruce Cornell, of CSIRO, and his colleagues, through the CRC for Molecular Engineering and Technology, a joint venture of the CSIRO, the University of Sydney and the Australian Membrane and Biotechnology Research Institute (AMBRI). Through AMBRI, the biosensor is now being developed for medical diagnostics.

We humans are not the only creatures to benefit from Australian biotechnology. The Animal Gene Storage Resource Centre of Australia (AGSRCA) is contributing to the conservation of endangered animals. AGSRCA—a cooperative venture between the Zoological Parks Board of NSW, Taronga Zoo, The Western Plains Zoo and Monash University—collects and stores samples of semen, embryos, testes, ovaries, skin and muscle from living or dead endangered animals. Material in the 'Gene Bank' is used for DNA studies and disease investigations, and semen and embryos can be transferred to animals in other regions, to maintain a healthy genetic diversity within the species.

AGSRCA scientists have also developed hormone tests to determine the optimum time to bring captive pairs of animals together for breeding, and are using new reproductive technologies developed for humans, such as in vitro fertilisation, to help preserve animal species.

That seems only right, for after exploiting nature's gifts to build a nation, it makes economic and environmental sense to return the favour.

BIOTECHNOLOGY

WHERE LAW AND SCIENCE MEET: THE SUCCESSFUL COMMERCIALISATION OF BIOTA'S SECOND GENERATION ANTI-INFLUENZA DRUG

BIOTA, a local biotech industry pioneer, and law firm ALLENS ARTHUR ROBINSON, which has specialist expertise in biotechnology law and patents, are working together on commercialising Biota's second-generation influenza drug.

The current high level of global commercial interest in Australian biotechnology reflects the remarkable achievements of companies like Biota. The creator of the world's first anti-flu drug, Relenza®, Biota continues to lead the world with the development of Relenza II, the next generation anti-influenza drug. Like its predecessor, the new treatment is inhaled and immediately begins to inhibit the spread of the virus within the lungs.

Relenza II will offer flu victims and their doctors an effective, safer and more easily administrable option. When Biota recognised the potential of the original drug, Relenza, planning for its successor began. As soon as Biota was ready to harvest the investment in Relenza II development, the company turned to the leading Australian law firm, Allens Arthur Robinson, to collaborate in the commercialisation process.

With many years of experience in working with both Australian and international biotechnology companies, Allens Arthur Robinson worked closely with Biota's technical team to secure leadership in the estimated US$1 billion worldwide anti-flu market.

This synergistic combination of technical, commercial and legal know-how ensured that the negotiations were underpinned by a complete understanding of the product as well as of Biota's commercial objectives.

The result? A heads of agreement contract with global pharmaceutical giant GlaxoSmithKline for the worldwide development and marketing rights that will see Biota continue to enjoy a significant revenue stream from Relenza II twenty years from now.

®Relenza is a registered trademark of the GlaxoSmithKline group of companies.

>> www.arh.com.au
and turn to page 170 and 173 for directory details

Hoskyn's Islands and reef, part of Queensland's Great Barrier Reef

MARINE SCIENCE

PICTURE: GREAT BARRIER REEF MARINE PARK AUTHORITY

MARINE SCIENCE

AUSTRALIA'S HUGE OCEAN TERRITORY PROVIDES US WITH BOTH OPPORTUNITIES ANS RESPONSIBILITIES, AND OUR MARINE SCIENTISTS ARE WORKING TO DO JUSTICE TO BOTH, REPORTS AMARA BAINS

Following the declaration of the United Nations Convention on the Law of the Sea (UNCLOS) in 1994, Australia became the guardian of one of the world's largest ocean territories. In return, Australia is required to further advance the knowledge of the territory's Exclusive Economic Zone (EEZ) through exploration and marine research and to develop, preserve and protect the marine zone.

The Australian Institute for Marine Science (AIMS), CSIRO Marine Science and James Cook University are among Australia's most highly regarded research organisations in marine science. These institutes share many international and national research affiliations, with a number of locally developed techniques finding application in countries outside Australia.

Marine science research is extremely important for Australia, as the marine sector has high conservation value and is worth some $35 billion per year to the economy—and it is growing at an annual rate of approximately 8 per cent. Tourism and aquaculture are major components of this sector, with tourism around the Great Barrier Reef worth billions to Australia's economy. The value of this is likely to grow over time as other reefs around the world face increasing pressure from human activities.

AUSTRALIAN SCIENCE

> **ABOVE:** Offshore tourism pontoon with image from wave atlas (inset) showing Queensland coastline and the Great Barrier Reef (in purple) and wave heights (in metres) expected to be exceeded during an average 20-year period. Pontoon photo courtesy of Fantasea Cruises.

NEW WAVE ATLAS GUIDES DESIGN OF OFFSHORE TOURISM PONTOONS

Developing the world's first cyclone wave atlas was a recent highlight for the CRC REEF RESEARCH CENTRE. When integrated with pontoon guidelines, the atlas helped the Great Barrier Reef Marine Park Authority and the tourism industry achieve world's best practice in the construction and mooring of offshore bases for tourism. The wave atlas has also been used for geomorphological studies of coastal beach profiles, assessing the risk of future cyclones to the Cairns region and by marine researchers to query the impacts of cyclones on coral reef growth.

The CRC Reef Research Centre is a knowledge-based partnership of coral reef managers, researchers and industry. It provides scientific information, education and training to enhance industries based on the reef, and maintain ecosystem quality and ensure sound management of the Great Barrier Reef World Heritage Area and other coral reefs around the world. CRC Reef Research Centre was established under the Australian Government's Cooperative Research Centres Program.

>> www.reef.crc.org.au
and turn to page 185 for directory details

There are many current research programs aimed at maintaining the integrity of our marine environment. The focus of these programs may be broadly categorised into the following areas: industry-based research (such as aquaculture, marine biotechnology, multiple use management of the EEZ and tourism); research on the impact of humans on water systems; exploration and conservation of marine biodiversity; and research into oceans and climate.

These divisions of marine research have largely arisen from consultation with industry and government bodies, and as a result, collaboration between the research organisations and industry is commonplace. One example of such a collaboration is the Reef Cooperative Research Centre, a joint research and development venture between the tourist operators of the Great Barrier Reef, who contribute a significant proportion of funding, the AIMS and other agencies. The role of this joint venture is to ensure the protection and long-term viability of the Great Barrier Reef and its associated tourism industry.

Another area of Australian industry experiencing rapid growth is aquaculture. A strong export market, especially in Asia, and a gross value of almost $400 million annually provide the impetus for focusing on farming techniques. A variety of aquatic species can be produced in Australia because of the climatic differences across the country: more than 60 aquatic species are now being farmed here, including pearl oysters, oysters, mussels, barramundi, prawns, salmonids, crayfish and algae.

MARINE SCIENCE

In addition to the development of breeding stocks, other research programs are focused on sustainable and improved nutrition, and on methods of protecting the aquaculture industry from disease outbreaks.

Research in aquaculture is carried out by a collaboration between the AIMS and many divisions of the CSIRO. Partnerships have been formed with industry members, the Fisheries Research and Development Corporation, Cooperative Research Centres (CRCs) for Aquaculture and Antarctic and Southern Ocean Research and the Australian Centre for International Agricultural Research, as well as with state and Commonwealth research bodies.

A major success of this collaboration is the development of a new pond management system, which has placed the $50 million Australian farmed prawn industry at the leading edge of environmentally friendly prawn farming. Part of the prawn pond management system is software known as Pondman. Adopted by most Australian prawn farmers as a management tool, the software provides a means of collecting and storing data on prawn pond characteristics. This data collection allows farmers to better understand what is working well in terms of prawn farm efficacy, and to make any necessary adjustments to maintain the productivity of their farm.

The AIMS has described all the coral species and mapped the length and breadth of the Great Barrier Reef, allowing for successful management of the area by the Great Barrier Reef Authority for the past 20 years. The monitoring program covers

PICTURE: CRC FOR REEF RESEARCH
> **ABOVE:** Crown-of-Thorns Starfish, which is a great danger to coral reefs

60 reefs per year and recently demonstrated its effectiveness by detecting the latest Crown-of-Thorns Starfish outbreak several years before the starfish actually reached the tourism-sensitive areas. The AIMS' knowledge of the coastal and reef processes and of marine communities in tropical Australia is of the highest calibre, and has often been applied to other areas of the tropical world.

The School of Marine Biology at James Cook University, both independently and through collaborations with the AIMS, has also contributed significantly to research on the Great Barrier Reef and problems in tropical marine science.

Biotechnology is a growing area of scientific research around the world, and the Australian marine environment may provide genetic resources for the development of pharmaceuticals, agrochemicals and bioremediation products. The AIMS has a

AUSTRALIAN SCIENCE

group within its structure, Marine Bioproducts, specifically devoted to the exploration of marine resources. Strategically located on the both the east and west coasts of Australia, Marine Bioproducts uses its expertise and access to a very diverse region to discover compounds for development by industrial partners. Currently the number of micro- and macro-organisms in their collection available for screening exceeds 10,000. The collection is screened for active compounds and enzymes using innovative approaches. The structure and function of active compounds are defined before further development by industry. Discoveries made in this process also contribute to our understanding of the role biochemistry plays in the marine environment and the complexity of the ecological interactions within marine communities.

discovered an ultraviolet blocking compound in reef corals and then developed and patented a synthetic copy of the compound. 'We have copied and modified nature's own defensive product, and have come up with a stable and efficient sunscreen that is not irritating to use, as it has low allergenic activity,' said Dr Walt Dunlap, one of the scientists involved in the project.

The CSIRO Microalgae Research Centre also conducts research in biotechnology, specifically, as its name suggests, in microalgae. The Centre isolates and cultures strains of microalgae to develop feed for aquaculture and new pharmaceutical and health products. Research is also conducted into algal blooms in order to understand their dynamics and to develop techniques for toxin control.

DISCOVERY OF BIOACTIVE PRODUCTS IS ONLY THE FIRST STEP IN TRANSLATING THE POTENTIAL WEALTH OF AUSTRALIA'S MARINE RESOURCES INTO THE REALITY OF COMMERCIAL PRODUCTS

Discovery of bioactive products is only the first step in translating the potential wealth of Australia's marine resources into the reality of commercial products. In pharmaceuticals, the AIMS has joined up with AMRAD, a Melbourne biotech research and development company, and a network of other Australian companies, all committed to the commercialisation of Australia's biomedical research; AMRAD is funding a major research project to discover new medicinal drugs.

The AIMS and an Australian company, Sunscreen Technologies P/L, are also developing the world's first natural marine sunscreen. AIMS

Australia's marine science research is recognised internationally. Recently awarded an IBM Environmental Research Program—one of only four to be granted worldwide—the AIMS received essential super-computer technology and funding to develop models incorporating the Institute's extensive historical field data with sophisticated 3D models.

In addition, methods developed by the AIMS have been used around the world, and a manual describing resource assessment protocols developed with Association of South-East Asian Nations (ASEAN) researchers has been recommended as an important reference on marine survey techniques.

MARINE SCIENCE

A partnership with US company Applied Science Associates (ASA) saw the combination of the AIMS' mathematical models of ocean circulation with the ASA user-friendly computer interface used to develop Oilmap, which allows prediction of the likely path an oil spill may take. The Australian maritime safety authority uses Oilmap as the national oil spill management system. Many oil companies around the Australian coastline use it to forecast the direction and fate of oil spills and for training personnel in oil spill emergency response and contingency planning.

The research conducted by the CSIRO on estimating ocean currents is solving a costly problem for the offshore oil and gas industry by providing a safe and economical means of developing offshore structures and submarine pipelines. A major offshore facility such as a platform or pipeline costs around $1 billion. The structure must be designed to withstand any storms likely to be encountered in its lifetime. Both under-design and over-design can lead to problems, ranging from very expensive repair work to catastrophic failure or substantial unnecessary expenditure.

Computer software developed by the CSIRO for modelling currents caused by tropical cyclones and tides has been used to predict ocean currents for the proposed West Australian Petroleum Ltd (WAPET) Gorgon platform and the Woodside second pipeline, which runs from the North Rankin platform, in Australia's northwest, to shore. The project is the latest application of 15 years of model development, data collection and interpretation within CSIRO Marine Research, and has applications extending beyond the offshore oil and gas industry.

PICTURE: GREAT BARRIER REEF MARINE PARK AUTHORITY

> **ABOVE:** Wave action on a reef crest

As marine science enters the new millennium, climate and global warming are becoming growth areas in research. The impact of global warming on coral reefs, and the impact of coastal development on land–sea interactions, are of major concern. There are significant research strategies in many organisations dedicated to these areas, as well as to others: developing biomarkers to detect sublethal biostresses caused by marine pollution; mapping seabed diversity; biotechnology and biodiversity. Such research strategies will ensure that Australian marine scientists and their advances will remain among the best in the world.

ECOLOGY AND THE ENVIRONMENT

AUSTRALIA'S UNIQUE ENVIRONMENT HAS FOCUSED OUR ECOLOGISTS' WORK ON SOLAR ENERGY, WATER AND SALINITY, AND THE PRESERVATION OF BIODIVERSITY, REPORTS LEIGH DAYTON

Around 160 million years ago, Australia was part of the great super-continent of Gondwana. Gradually, Gondwana began to break apart. The landmasses we know as South America, Africa and Madagascar, Antarctica, and Australia and New Zealand took shape. And by 45 million years ago, Australia was on its own, drifting slowly northwards with a cargo of plants and animals.

Today, Australian scientists are working to protect this unique and fragile continent. The lessons they learn will benefit the living land, in Australia and abroad. Better environmental management and planet-friendly technologies will reduce the ever-increasing impact of human activity on our world.

Already, Australian teams made up of people from academia, industry and national research bodies lead the world in the development of renewable energy and other clean, green technologies. Many of the new technologies come from a creative blend of fundamental science and novel ways of viewing the everyday world.

AUSTRALIAN SCIENCE

Created in the laboratory, the miniature world of nanotechnology—the manipulation of single molecules and atoms to create tiny tools and processes—promises big gains for scientists with CSIRO Telecommunications and Industrial Physics. Dr Vijoleta Braach-Maksvytis and her team, for example, have devised silicon computer chips that build themselves. The self-assembly chips could replace existing, highly toxic microchip fabrication methods.

The group is also at work on artificial photo-synthesis, a procedure that mimics the way plants convert sunbeams into energy. The technology could be used to clean up carbon dioxide waste emitted during industrial processes. The by-product could be turned into an alternative source of fuel.

Australia's annual water consumption, always a matter for concern in this dry continent, has been drastically reduced with the use of dual flush toilet cisterns, which were introduced in 1980. Users have a choice of a full 9.0 litre flush or a skimpy 4.5 litre half-flush. And once the 'flush' hits the sewerage system, it can now travel down high-density polythene pipe produced from recycled plastic bottles, such as the Ribloc Expanda Pipe, which is now licensed in over 20 countries.

Finally, the flushed sewage can be treated with the Memcor continuous microfiltration process. Here, hollow plastic fibres, dotted with millions of tiny pores, filter out waste particles as small as a disease-causing virus. This technology has been bought by Vivendi Water, the most comprehensive water treatment and supply company in the world, which delivers services ranging from home filtration systems to the construction and operation of commercial waste water treatment plants.

Some of the most important environmental advances made by Australian scientists come from their use of the famous qualities of the island continent: it's big, flat, dry and drenched in sunlight.

Professor Martin Green, of the Centre for Photovoltaic Engineering at the University of

> **SOME OF THE** MOST IMPORTANT ENVIRONMENTAL ADVANCES MADE BY OUR SCIENTISTS COME FROM THEIR USE OF AUSTRALIA'S FAMOUS QUALITIES: IT'S BIG, FLAT, DRY AND DRENCHED IN SUNLIGHT

New South Wales (UNSW), leads important solar cell research which is helping to 'sell' solar power to budget-conscious governments, industries and citizens worldwide. His team has set the pace, producing the world's most efficient silicon solar cell. The power pack converts 24.7 per cent of the sunlight that hits it into electricity.

These 'first-generation' solar wafers are now complemented by a second generation of thin-film solar cells, which promise to reduce costs by 90 per cent. Right now, UNSW researchers are planning a third generation of photovoltaic technology. The idea is to increase efficiency by stacking cells, one above the other.

The Centre has forged links with domestic and overseas industry, aiming to make solar power cheap enough to compete economically with power

> **ABOVE:** Dish-shaped solar power reflectors at the solar power station at White Cliffs, New South Wales

generated by fossil fuels, which release carbon dioxide and other pollutants into the atmosphere.

Collaborative research is underway with two Australian photovoltaic manufacturers, Pacific Solar and BP Solar, the world's largest solar cell manufacturer. Australian-manufactured solar cells were prominently displayed in the 655 houses of the athletes' village at the Sydney 2000 Olympic Games.

Overseas, the European Commission has provided funding to the Centre to support partners in England and Spain. The Italian group Eurosolare has licensed technology developed at the Centre, and is working on research with the Australians.

Scientists at the University of Sydney are also taking Australian solar science to market. Under the leadership of physicist Dr David Mills, University staff are working on Australia's largest solar power plant project. Scheduled for commissioning in late 2001, the 4 megawatt solar thermal power plant will supply clean energy to an existing coal-fired plant. It is poised to be the lowest-cost solar electricity plant anywhere.

David Mills has also pioneered the Solar City program—its goal is to encourage key cities around the globe 'to adopt a target of sustainable emissions' by 2050. The International Energy Agency has adopted the scientific part of the program, now headed by University of Sydney urban design professor Peter Droege. Cities in Scandinavia, Germany, Italy and Mexico, as well as China, South Korea and Australia, have expressed strong interest in the program, slated to begin in 2001.

AUSTRALIAN SCIENCE

Meanwhile, wind power projects are gearing up across the nation. CSIRO scientists created a computer modelling technique that allows them to pinpoint the areas of richest wind energy potential in a given geographic area. Working with NSW energy supplier Pacific Power, a team from CSIRO Land and Water have identified a major wind 'hot spot' near Crookwell, in the Southern Tablelands of New South Wales. The site is to become the home of Australia's first large-scale wind farm, and will be directly connected to the power grid. With a good wind blowing, the Crookwell farm will produce 5 megawatts from eight propeller-driven turbines, enough to meet the average electricity demand of at least 3,500 homes. It will reduce greenhouse gas emissions by 8,000 tonnes per year.

a virus that kills only rabbits. RCD has proven effective, especially in arid and semi-arid areas. It is now used in conjunction with the less lethal myxoma virus, a rabbit-killing virus discovered and developed by the CSIRO in the 1950s.

Together, these viruses have cut the rabbit population dramatically. Consequently, dozens of species of mulga scrub, acacia trees, and other native plants have regenerated, helping to revitalise the ecosystem. A modified version of RCD is under development in Western Europe, where wild rabbits are now causing environmental degradation.

Several other outstanding projects are exploring inexpensive methods of assessing and repairing land and water, damaged by 200 years of inappropriate farming techniques and land clearing.

SEVERAL OUTSTANDING PROJECTS ARE EXPLORING INEXPENSIVE METHODS OF ASSESSING AND REPAIRING LAND AND WATER, DAMAGED BY 200 YEARS OF INAPPROPRIATE FARMING TECHNIQUES

Beyond technology, Australian scientists have joined forces to tackle biological and ecological problems. They are aiming to find ways of reintroducing endangered animals to the environment and ridding the continent of animal pests and weeds.

Their teamwork has paid off. Starting with a hunch from biologist Dr Brian Cooke, of CSIRO Wildlife and Ecology, CSIRO researchers led the decade-long Australian and New Zealand effort to develop a new biological weapon against European rabbits: since their introduction in the 19th century, rabbits have devastated native plants and caused massive soil erosion across Australia. The result of the team's work is rabbit calicivirus disease (RCD),

University researchers and groups such as the Cooperative Research Centre (CRC) for Dryland Salinity and CSIRO Land and Water are exploring water and farming technologies that will reverse the buildup of salt in waters and soils. The problem is greatest in the large irrigation areas of the Murray–Darling Basin, set up at the turn of the century. In other regions, such as the dryland agricultural areas of central and northern New South Wales, salinity is a more recent phenomenon.

Without intervention, the annual movement of salt in the landscape will double in the next hundred years across all major valleys—three to five million hectares of land will suffer serious salinisation within 50 years, and wetland and river salinity will climb steeply.

To meet the challenge, farmers are planting trees, creating shelter belts, retaining native vegetation, adopting drip and trickle irrigation, practising minimum tillage and reducing the use of chemicals. Their efforts are bolstered by the Landcare program, which aims to plant one billion trees.

Dr Brian Loveys, of CSIRO Plant Industry, has discovered that grapes and some horticultural and citrus crops stay healthy if only part of their root system is watered. This 'partial root zone drying' technique will enable growers to conserve water.

The Arid Zone Recovery Project, another collaborative effort, is aimed at boosting native plants and animals in a region ravaged by poor land management. Scientists and concerned citizens in South Australia have successfully removed rabbits, stock and non-native predators, such as cats and foxes, from the 60 square kilometre project area at Roxby Downs, and recently reintroduced the Greater Bilby, Australia's 'Easter Bilby', to a 14 square kilometre enclosure. The little long-eared marsupial is an endangered species of bandicoot. Other native species, including Burrowing Bettongs and Greater Stick-nest Rats, have also been reintroduced to the area. Plans are underway to restore other species to their former habitat.

Another approach is Dr John Wamsley's 'Earth Sanctuaries' concept, begun in 1969 in the Adelaide Hills of South Australia. Damaged land is purchased and fenced to exclude introduced animals such as foxes, rabbits and cats, and once the land is

ECOLOGY AND THE ENVIRONMENT

REDUCING THE DEVASTATION CAUSED BY INTRODUCED PEST ANIMALS

THE COOPERATIVE RESEARCH CENTRE FOR THE BIOLOGICAL CONTROL OF PEST ANIMALS utilises biotechnology to achieve significant and continuing benefit for Australia by reducing the devastating environmental and economic impact of introduced pest animals.

Novel fertility control agents are being developed which are more cost-effective, environmentally friendly, more humane and more effective than current methods. The fertility control agents under development are based on immunocontraceptive vaccines delivered either orally, via a bait or through a virus which specifically infects the target pest animal population.

Research is currently focused on three of Australia's most pervasive pests: the European rabbit, the European red fox and the introduced house mouse. Other applications of the technology will include the control of other pest animals (such as the rat, cat and pig) and better strategic management of a number of emerging wildlife diseases which threaten human and livestock health.

>> **www.pestanimal.crc.org.au**
and turn to page 184 for directory details

AUSTRALIAN SCIENCE

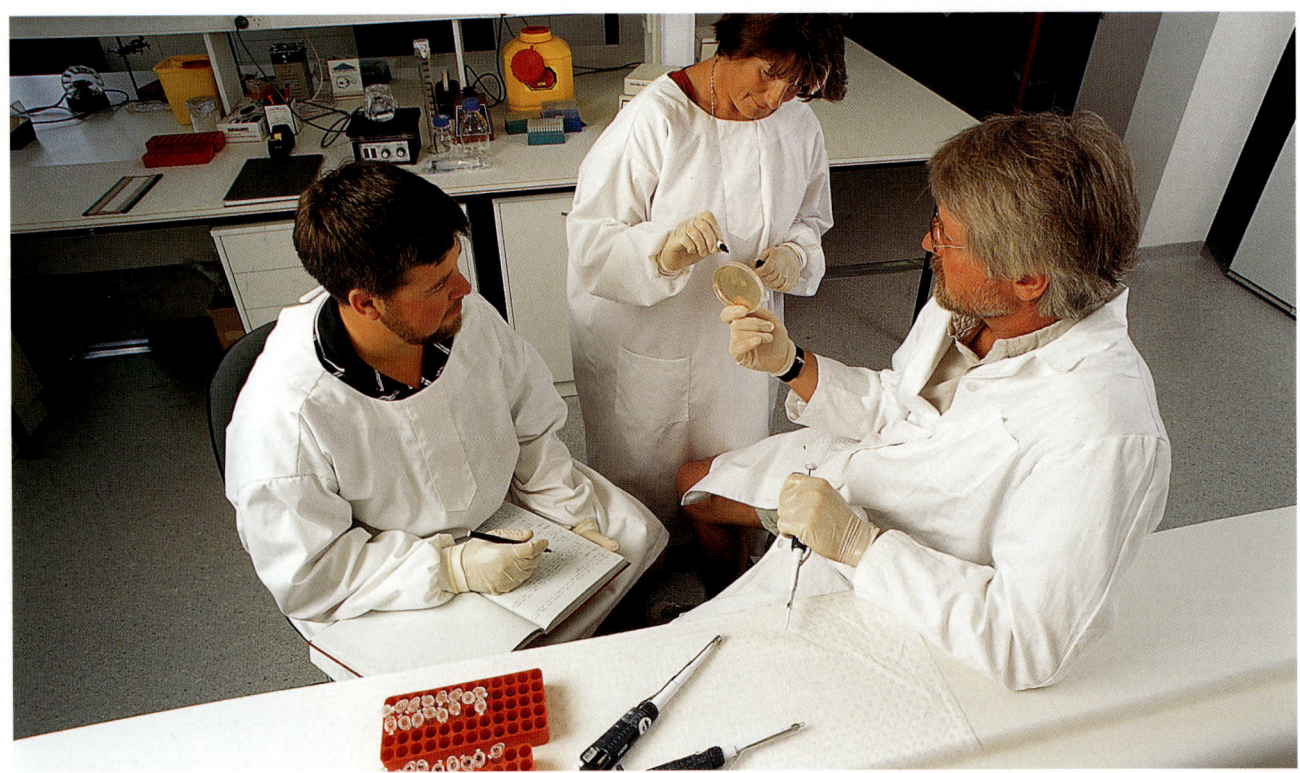

> **ABOVE:** Scientists are working on fertility control for pest animals, and on managing wildlife diseases which affect human and livestock health

PICTURE: CRC FOR THE BIOLOGICAL CONTROL OF PEST ANIMALS

rejuvenated, rare and endangered native animals are reintroduced. Earth Sancturies Limited is now a publicly listed conservation company, supporting four sanctuaries in South Australia and New South Wales. A fifth is poised to open in Victoria.

Many other reintroduction projects are in progress, but before experts can restore Australia's ecosystems to pre-European condition, they must find out what existed then. Surprisingly, while early naturalists marvelled at the exotic species here, from kangaroos to koalas, researchers are only now beginning to understand Australia's biodiversity, to tally its full complement of plants and animals.

Already, the numbers are impressive: 282 species of mammals, 750 of reptiles, nearly 800 of birds, 200 of frogs and 225,000 of invertebrates (animals without a backbone). At last count there were 140,000 species of insects. Most of these are found nowhere else on Earth. About 90 per cent of Australia's flowering plants are unique, as are half the bird species, over 80 per cent of the reptiles, and innumerable micro-organisms and insects.

Although Australia's plants and animals are unusual, many live in harsh, dry conditions that are common worldwide. In fact, most countries are arid or semi-arid. In contrast, most research on the way ecosystems work has been done in temperate lands, by researchers from North America and Europe. The result is a 'eurocentric' view of natural systems, a view that is skewed to temperate climes.

The implication is clear. What scientists discover about Australia—its wetlands, deserts, and flood plains, its salty lakes and claypans, its semi-permanent lakes and streams—will benefit not just Australians. It can provide insight into the ecology and management of other parts of the world.

And all this from a chunk of the globe that went walkabout millions of years ago.

ECOLOGY AND THE ENVIRONMENT

> **RIGHT:** Helping to restore important mangroves at Sydney's Homebush Bay—Dr Gee Chapman and Prof. Tony Underwood

THE IMPACT COASTAL DEVELOPMENT HAS ON ANIMALS AND PLANTS

THE CENTRE FOR RESEARCH ON ECOLOGICAL IMPACTS OF COASTAL CITIES provides the strategic information required to understand what cities do to animals and plants in coastal and estuarine habitats.

Novel sampling methods to detect environmental impacts in variable habitats have been developed by the centre's director, Professor Tony Underwood, and have led to an increased capacity to understand disturbances in coastal regions. The centre has also identified considerable small-scale patchiness in many coastal populations which has led to increased attention to the ways we measure natural changes over time.

Although animals and plants often live across wide ranges of habitats, most existing research does not integrate patterns and dynamics from one habitat to another and has been conducted on too small a scale. Much is known, for example, about the ecology of rocky shores, mangroves and seagrasses, but little is known about how these habitats interact.

To rectify this weakness in existing data, the centre has brought together some of the world's leading researchers in a range of related fields. The centre strives to integrate their ideas and models across habitats, understand the processes of recolonisation after disturbances, interpret the way populations persist when they are disturbed and predict the patterns of occupation of new habitats when they are built.

Its research focusses on the development of new methods to assess ecological changes and procedures for interpreting changes when they occur. Both require long-term, large-scale studies and the use of modern experimental techniques to test theories about responses to disturbances to improve predictions for future responses to environmental change.

By bringing together expertise from a range of ecological disciplines (including experimental design, statistical analysis, genetics, population processes and modelling), the centre has created a new focus for understanding natural changes in addition to those changes brought about by coastal development.

>> 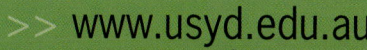 www.usyd.edu.au
and turn to page 182 for directory details

AGRICULTURE

Flowering switch gene in a common weed called *Arabidopsis thaliana* **PICTURES:** CSIRO PLANT INDUSTRY

AGRICULTURE

FROM COTTON AND WHEAT TO POTATOES AND SUGAR, AND FROM CATTLE TO POULTRY, AUSTRALIA'S AGRICULTURAL SCIENTISTS ARE PRODUCING WORLD-CLASS ADVANCES, AS GORDON COLLIE REPORTS

Australian agriculture is on the brink of a revolution, as science unlocks the genetic secrets of plants and animals. Gene technology research has become a key component of science projects—across the rural spectrum—which underpin industries generating export income of more than $25 billion a year. Advances in genetic science hold the promise of a quantum leap for the rural sector, bringing significant economic, social and environmental benefits.

Modified plants with inbuilt pest resistance that dramatically reduces the amount of chemical spraying needed, for example, are already a commercial reality. Researchers are working on a myriad improvements, from drought- and salinity-tolerant crops to plants with enhanced nutrition and health benefits.

Knowledge of livestock genetics is also advancing rapidly, identifying traits from superior fibre and animal growth rates to natural protection against diseases.

A genetic research breakthrough of global significance has been recognised with the inaugural Prime Minister's Prize for Science.

For Dr Jim Peacock and Dr Liz Dennis of CSIRO Plant Industry, 20 years of collaborative research led to the discovery of a key gene which controls flowering in plants—the Flowering Switch Gene. Isolation of this gene has the potential to greatly improve the performance of a range of food crops, as many crops risk huge yield losses through cold snaps or heatwave conditions at the critical flowering period. By manipulating the Flowering Switch Gene it will be possible to produce strains of crops that flower at the right time for the climate in which they are grown, reducing the risk of yield losses.

Discovery of the gene came from an investigation into the reasons why plants flower after being subjected to a period of low temperature. In experiments, the scientists were able to minimise the need for cold before flowering by manipulating the plant's DNA. Further work on a strain of plant that flowered very late led to isolating the gene that caused the late flowering mutation. Switching off the repression gene caused the plant to flower earlier than it normally would. This key piece of patented research has given scientists a better understanding of the flowering process, and is expected to have worldwide benefits for plant breeding.

The cotton industry has been one of the first to embrace gene technology, with many Australian farms now growing varieties that protect themselves against chewing insects by producing a natural pesticide in their leaves.

Cotton has become a major crop through New South Wales and Queensland, but the high level of spraying needed to control insects has raised environmental concerns, leading to research into ways in which plants that protect themselves could be bred. Scientists at Monsanto in the United States developed a gene—from a bacterium commonly found in the soil, *Bacillus thuringiensis*—which has been successfully adapted to cotton varieties suitable for Australian conditions, providing inbuilt protection for the cotton plant against chewing insects.

The area of modified cotton grown has increased to about 25 per cent of the total crop as farmers incorporate it into an integrated pest

THE COTTON INDUSTRY HAS BEEN ONE OF THE FIRST TO EMBRACE GENE TECHNOLOGY, WITH MANY AUSTRALIAN FARMS NOW GROWING VARIETIES THAT PROTECT THEMSELVES AGAINST CHEWING INSECTS BY PRODUCING A NATURAL PESTICIDE IN THEIR LEAVES

management strategy, reducing the need for pesticide sprays by 40 to 60 per cent.

Scientists in Australia have been working on the development of plants containing two protective genes for even more effective insect control; these are expected to be commercially available to farmers by the 2003 planting season.

To protect the new technology from the threat of insect resistance developing, use of the modified varieties is strategically managed, with increased emphasis also given to the important role played by beneficial insects, the natural predators of cotton-eating pests.

During the past few years, great progress has been made breeding high-yielding cotton varieties, and the fibre quality of Australian cotton is now rated among the best in the world.

Water has also been a major focus of cotton research, and technology that makes the most efficient use of a limited resource and protects

PICTURE: AVENTIS CROPSCIENCE PTY LTD

> **ABOVE:** Cotton is a growing industry in Australia, and research that has led to reduced pesticide use is increasing its sustainability

the environment by containing pesticide residues is now being adopted. Water balance science helps optimise crop yields while minimising water use, and also lowers the risk of soil salinity, a major concern for long-term sustainability.

In horticulture, major advances are being made with integrated pest management (IPM) across a number of crops. With the potential for big cost savings and environmental benefits, more and more farmers are adopting the technology.

The Horticulture Research and Development Corporation has recognised the work of Queensland Department of Primary Industries scientist Dan Smith by awarding him the Graham Gregory Award for his work in the state's citrus industry. More than 80 per cent of growers are now using IPM technology to manage problems of red scale and oriental mite with reduced chemical use. A range of parasitic wasps and predatory mites from overseas has been successfully introduced to combat exotic pests that have no natural enemies in Australia.

Genetic technology is being developed to confront one of the most serious problems in the potato industry—leaf roll virus. This destructive virus can cut crop yields in half and seriously downgrade potato quality. A gene from the leaf roll virus inserted into the potato's DNA has successfully resisted infection in trials, reducing pesticide use.

Research has also shown that brown or black spot bruising, which occurs during mechanical harvesting of potatoes and other crops, can be countered by modifying a plant enzyme gene.

In the sugar industry, important research into compiling a sugarcane gene bank has been completed. Australia's first sugarcane genomics project has resulted in a database containing the identities

AUSTRALIAN SCIENCE

> **ABOVE:** The CRC for Molecular Plant Breeding is a world-class research, education and training centre in the field of molecular plant breeding.

SCIENCE AND TECHNOLOGY SUPPORT PLANT BREEDERS

THE COOPERATIVE RESEARCH CENTRE FOR MOLECULAR PLANT BREEDING is focused on developing, testing and implementing effective strategies for cereal and pasture grass breeding programs and utilising the new technologies of molecular biology.

Its main areas of expertise are plant pathology, physiology, nutrition, stress tolerance and breeding combined with molecular biology, genetic engineering and cereal chemistry.

The techniques employed in genetic engineering and molecular markers are the most sophisticated tools currently available for genetic analysis and manipulation of crop plants.

The CRC's research results are enabling Australia's cereal and pasture grass improvement programs to remain internationally competitive and responsive to the changing demands of agricultural practice and consumers.

A close association with plant breeders in Australia and overseas, as well as strong links with international seed companies, is ensuring the rapid and effective implementation of new technologies and plants strains developed by the centre.

The CRC's strategies are providing a research base, tools and training for the next generation of plant breeders.

>> www.molecularplantbreeding.com
and turn to page 185 for directory details

of 4,000 genes expressed in cane stems. The entire gene bank has been documented and stored in duplicate as a resource for sugar researchers.

Another weapon has been added in the ongoing battle with cane grubs, the major insect pest affecting cane crops. The first biological control agent based on a naturally occurring soil fungus showed promising trial results ahead of its commercial release as a treatment for greyback grubs.

In the grains industry, research effort has concentrated on issues affecting access to international markets for a wide range of crops outside the staples of wheat and barley. Grains such as sorghum, oats, maize and triticale, pulses such as lentils, field peas and chickpeas, and oilseeds such as canola, sunflower and soybeans have all benefited from a greater understanding of the requirements of overseas customers.

With around one-third of Australia's wheat exports used for noodle production, scientists are helping the industry stay ahead of its competitors. Noodle colour is an important attribute, and researchers have found that it can be influenced by the level of a key plant enzyme. Screening for this trait in breeding programs could substantially improve noodle colour. Gene technology is also being used to help identify wheat lines that have the starch characteristics the Japanese noodle market seeks. This research is paying big dividends for the grain industry, as the best noodle wheats command premium prices.

Gene technology advances in the livestock industry have reached the point where DNA-based products are now being released into the commercial market.

AGRICULTURE

PICTURE: AVENTIS CROPSCIENCE PTY LTD

> **ABOVE:** Barley—research into the entire growing process, from seed selection to harvest, is benefiting a range of grain crops in Australia

The launch of GeneSTAR Marbling, for instance, heralded a new era for cattle breeding—it was the first commercial DNA diagnostic test in the world for a beef production trait. The highly accurate test can be made on a sample of an animal's hair roots, semen or blood, and will determine the presence of a key gene which influences fat marbling, a desirable feature in beef destined for markets such as Japan. Animals can carry none, one or two copies of the gene.

The technology, which is based on original CSIRO research, has been patented internationally, and the meat marbling gene test has been brought to the market by the company Genetic Solutions, which is developing a range of other tests to assist cattle breeders. These tests will make it easier and cheaper to detect traits such as disease resistance and meat tenderness, for example, which are currently difficult and expensive to measure.

The company has also developed SureTRAK, a genetic testing system used to ensure that meat is traceable as it moves along the production chain. This provides a product integrity guarantee from paddock to plate, based on a DNA sample of each animal being collected and archived.

Australia enjoys an international reputation for excellence for the animal disease diagnosis and research carried out at the Australian Animal Health Laboratory (AAHL) at Geelong, in Victoria.

AUSTRALIA ENJOYS AN INTERNATIONAL REPUTATION FOR EXCELLENCE FOR ANIMAL DISEASE DIAGNOSIS AND RESEARCH CARRIED OUT AT THE AUSTRALIAN ANIMAL HEALTH LABORATORY (AAHL)

AUSTRALIAN SCIENCE

This facility is recognised as one of the best in the world for the safe handling and containment of infectious micro-organisms.

The Laboratory, now part of the newly structured CSIRO Livestock Industries, will continue its pivotal role in protecting animal health. Research capability in this field will be boosted by the location of a sister laboratory within the Queensland Institute of Molecular Biosciences—this is due to open at Queensland University in mid-2002.

AAHL has successfully tackled a succession of animal health problems, including black disease and cheesy gland in sheep and pleuropneumonia and tuberculosis in cattle. Eradication or control of these and other diseases yield ongoing benefits for livestock producers.

The Laboratory has also played a key role in combating several outbreaks of the deadly Newcastle disease, which strikes the poultry industry. A national survey of the disease has been undertaken, with the laboratory characterising and pathotyping viruses being sent in from around the country.

Also, a search is underway for a faster, more reliable test for Johne's disease, a wasting disease of sheep and cattle. Scientists are aiming to produce a diagnostic test that accurately shows whether or not an individual animal has the disease.

Attempts are also being made to develop new blood tests to help protect Australia from Japanese encephalitis, a serious viral disease of humans and horses. Disease outbreaks have occurred in the Torres Strait, to the near north of Australia, with one reported human case on Cape York Peninsula.

Pig and cattle blood samples from northern Australia and Papua New Guinea are routinely tested as part of a quarantine early warning system.

Scientists at AAHL have two promising research projects on small proteins known as bacteriocins, which could be used as an alternative to antibiotics for intensively housed chickens and pigs. The use of these proteins as a medication would help reduce the risk of human pathogens becoming resistant to antibiotics. These proteins also quickly break down in animals, so there are no residues left in meat.

Major collaborative dairy industry research has probed the health benefits of milk proteins, including the potential to protect against cancers. Work has concentrated on whey, an often underrated dairy by-product, and the influence of dietary protein on the development of colon and skin cancer. Trials have shown that whey protein, either as a total powder or isolated as individual fractions, offers protection against both forms of cancer. While the health value of whey has long been recognised, it has not before been subjected to rigorous scientific scrutiny. Other trial work has examined the influence of protein on the functioning of the immune system, and the efficacy of whey growth factor extract in wound healing and gut repair.

Several commercial yoghurts and cheeses have also been analysed for their peptides, which have potential health benefits. Health-giving foods are expected to become a lucrative sector of the food market in the future, and this research is seen as an essential foundation for dairy companies' plans to take advantage of these commercial opportunities.

AGRICULTURE

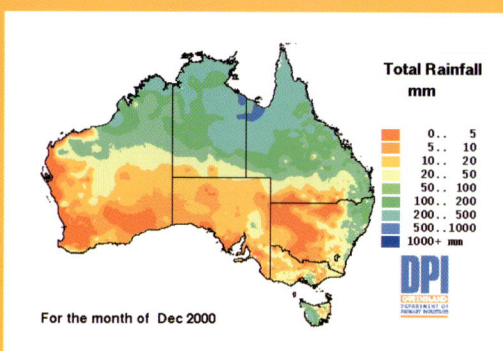
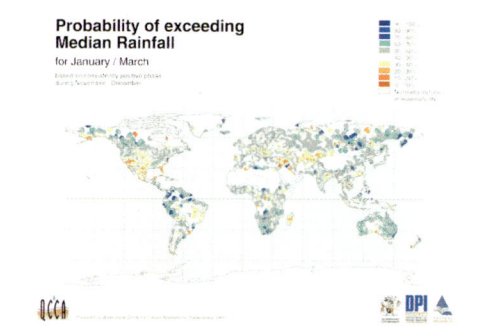

> **ABOVE:** Snapshots from the **www.dnr.qld.gov.au/longpdk/** website showing rainfall patterns throughout Australia
> **RIGHT:** Radiosonde weather balloon for information on high-altitude weather, Giles, WA

>> WEATHER SCIENCE

QUEENSLAND CENTRE LEADS THE WAY IN CLIMATE RESEARCH, CLIMATE PREDICTION AND CLIMATOLOGY TRAINING

World-class research that has a vital impact on rural industry is conducted through the Queensland Centre for Climate Applications. The Queensland state government sponsored the establishment of the centre to help producers cope with the most variable weather in Australia.

Scientists have undertaken climate forecasting research, and can now predict seasonal changes up to six months in advance—this is a valuable tool in farm management planning.

Skills developed at the centre are now in demand around the rest of Australia and internationally, and Australian weather experts have consulted in South America, Southeast Asia, southern Africa and the Indian subcontinent.

An extensive collection of weather data that allows property owners to customise their own forecasts is published on a website, called The Long Paddock (**www.dnr.qld.gov.au/longpdk/**). An allied project, Aussie GRASS (**www.dnr.qld.gov.au/longpdk/agrass**), combines weather and natural resource data in simulated models of pasture growth on a 5 km grid over the entire state.

The Centre has also demonstrated a commitment to the future of climate research by combining with the University of Southern Queensland to offer the world's first undergraduate course in climatology. The degree has aroused strong overseas interest and is expected to address a global shortage of experienced professional climatologists.

Stacked washed coal, Utah mine, Queensland
> RIGHT: Workers on offshore gas processing platform

MINING AND RESOURCES

MINING AND RESOURCES

MINING AND RESOURCES ARE AS IMPORTANT TO AUSTRALIA NOW AS THEY EVER WERE, AND RESEARCH CONTINUES TO OPEN UP NEW FIELDS AND IMPROVE MINING PROCESSES, REPORTS GORDON COLLIE

From the crude extraction of shallow coal for domestic fires in the Sydney penal colony more than 200 years ago, mining and minerals development have played a crucial role in shaping a modern and prosperous Australia.

Gold rush fever helped the nation achieve a critical population mass, sparking a multicultural boom as settlers arrived from around the world seeking their fortune in a harsh new land. And the birth of an industrialised nation can be traced to the chance discovery of massive silver, lead and zinc deposits in remote Broken Hill (New South Wales) in the 1880s.

The settlement and mining at Broken Hill triggered the mining industry research which has underpinned national development for more than a century. Smelting and processing technology pioneered there has been widely adopted around the world.

Mining science in the modern era continues the tradition of excellence, with hundreds of millions of dollars now invested in the ongoing quest for better technology. This investment in research has been richly rewarded, with Australian minerals and energy exports to the world now worth more than $50 billion a year.

Ausmelt, for example, based in Dandeong, Victoria, is commercialising the Sirosmelt Top Submerged Lance (TSL) smelting method

AUSTRALIAN SCIENCE

invented by Dr John Floyd at the CSIRO. TSL is a low-cost, high-intensity system for smelting base metal ores and concentrates, as well as for recovering high value from wastes. Smelting is rapid and furnace residence times are minimal, which means significantly lower capital and operating costs than alternative technologies. The company now has a global business in the non-ferrous metal industry, with plants built or being built all over the world.

While the CSIRO has been the flagship for mineral and resource science in Australia for decades, the research base is now broad, with a mix of university, government and private institutions drawing in significant private industry resources as well as public funding. A host of specialised Cooperative Research Centres (CRCs) foster collaborative research which is achieving strong industry support because it is focused on achieving commercial outcomes.

With the relative decline in mining and resource development in both Europe and the United States, Australia and Canada now stand out in terms of their research capabilities. Australia's major mining and resource states, Queensland and Western Australia, are home to a number of mineral science establishments, and the research environment there is drawing increasing international respect, and winning accolades across the spectrum—from exploration and mine planning to the development of software for use in high-technology processing.

In a bold example of collaborative research excellence, scientists in locations all around Australia are pooling knowledge for an ambitious project known as the Glass Earth concept—the aim is to make the top one kilometre of the Australian continent as transparent as glass. Scientists plan to create a virtual map of the bedrock deep below the surface layers of the Earth, thus increasing our understanding of how these giant ore bodies were created and using this knowledge to predict where new ones can be found. To do this, a powerful suite of new technologies for aircraft, satellites and ground equipment will be used to scan the landscape in the visual and infrared spectra, and for electromagnetic signals.

Looking below the surface of the Earth and learning how the continent's minerals have been rearranged over the past 150 million or more years presents a unique set of challenges in physics, chemistry and hydrogeology, and requires the ability to correlate massive datasets. Such a revolutionary approach to mineral exploration is expected to dramatically reduce costs and give the Australian industry an edge in increasingly competitive world markets.

Mineral science infrastructure is also receiving a major boost, with the expansion of the Queensland Centre for Advanced Technologies (QCAT) in Brisbane and the construction of the new Australian Resources Research Centre in Perth.

AUSTRALIA'S RESEARCH ENVIRONMENT IS DRAWING INCREASING INTERNATIONAL ATTENTION AND RESPECT, AND WINNING ACCOLADES ACROSS THE SPECTRUM

MINING AND RESOURCES

> **ABOVE LEFT:** Field Emission SEM image of char particle formed under pressure.
> **ABOVE:** Ash deposition in ACIRL's combustion test furnace produced in a blended coal burn for the Centre.

MAXIMISING THE VALUE AND ENVIRONMENTAL PERFORMANCE OF AUSTRALIAN BLACK COALS

THE COOPERATIVE RESEARCH CENTRE FOR BLACK COAL UTILISATION is the first sustained collaborative research venture between Australia's black coal producing and using industries. Consisting of four research and eleven industry participating organisations, its research objectives are embodied in its mission statement: *To maximise the value and environmental performance of Australian black coals.*

The Centre was instrumental in saving a long-term contract worth $400 million to a participant company by co-ordinating input from a range of company technologists, consultants, university researchers and international customers.

Modern techniques such as scanning electron microscopy are applied to characterising coal and coal ash. This enables the behaviour of ash to be understood from its origins in each pulverised coal particle. The application of such knowledge brings major benefit to both the coal and electricity generating industries.

The Centre also provides a world-class set of databases for ash minerals and their reactions, an achievement made possible through its recognition of the synergy gained by incorporating metallurgical and material science into coal research. One way in which this knowledge can be used industrially is for product quality control of minesite coal blending operations to meet ash quality specification. Participant electricity generators have also been able to minimise costs and protect asset life using tools provided by the Centre.

International recognition of the Centre's research capabilities is highlighted by external research contracts worth over $1.5 million to date. An international research review panel recently acknowledged the Centre's world-class research by declaring its mineral matter reactions program to be: *'leading the world in this field'*. Similar praise came from Ishikawajima-Harima Heavy Industries, Japan's leading power generation industry boiler manufacturer: *'We see the Black Coal CRC as world leaders in this type of work'*.

Such acclaim is testament to the Centre's focus on research quality and relevance.

This partnership in quality research started by the Black Coal CRC will be continued in the successor, The Centre for Coal in Sustainable Development (CCSD), from 2001 to 2008.

>> www.newcastle.edu.au/department/black_coal_crc
and turn to page 174 or 184 for directory details

AUSTRALIAN SCIENCE

> **ABOVE:** Conveyor belt mining machinery at Elura lead and zinc mine, western New South Wales

The expanded (stage two) QCAT is a state-of-the-art joint venture between the CSIRO and the Queensland government, and brings together a unique mix of disciplines—the CSIRO's divisions of Exploration and Mining, Minerals, Energy Technology and Manufacturing Science and Technology, and the CRCs for Black Coal Utilisation, Cast Metals Manufacturing, Landscape Evolution and Mineral Exploration and Mining Technology and Equipment—to focus on mining and resource technology. The just-completed stage two is a phased expansion of the original site in Brisbane's western suburbs, and includes a new research laboratory, a technology transfer building and expanded research support facilities.

Coal is Australia's largest export industry, with overseas earnings worth almost $9 billion annually . The coal technology used in Australia is world class, but research scientists are continually striving for improvements in all facets of coal production.

CSIRO Energy Technology and the Australian Coal Association Research Program have developed a new concept for fine coal cleaning—using enhanced froth flotation—which delivers significant benefits to industry. The patented process, known as TurboFlotation, can accomplish separation at fine sizes in a matter of seconds, compared with several minutes for conventional flotation. A 90 mm pilot scale separator operating at 20 times the volumetric capacity of conventional systems has led to scale-up plans for a one metre diameter unit. This commercial-scale unit will be capable of treating fine coal at flow rates of up to 600 cubic metres an hour. Because the technology can achieve a substantial expansion in throughput, a much smaller infrastructure is required, opening up the potential for major savings on capital as well as on operational costs. Numerous other applications for this technology have been identified in the mineral and waste treatment industries.

A number of outstanding projects are being advanced through the CRC for Mining Technology and Equipment, including a revolutionary cutting technology which will have application in both mining and civil engineering projects. Conventional disc cutters use very high compressive forces to fracture rock and need very large support machines. The new oscillating disc cutter, however, is a light-weight, highly manoeuvrable continuous excavation machine capable of achieving high production rates in strong rock.

Four innovative concepts have been combined in the new technology, including undercutting to break the rock tension and the use of water jets to assist the cutting action and provide cooling. Cutting discs up to 100 mm in diameter have been successfully tested in the laboratory and, with the aid of mathematical modelling of the rock cutting action, a 300 mm diameter prototype has been developed.

Another project with significant potential involves the development of an ultra-short radius drilling system that will enable methane gas from multiple coal seams to be effectively and economically drained from the surface.

The technology could be used either to reduce seam gas content ahead of underground mining, which would enhance operational safety, or as a method of large-scale commercial gas extraction. Australia's vast resource of coal-bed methane has the potential to meet expected energy demands for many years, but existing extraction methods make the methane too expensive for commercial exploitation.

Based on the outcomes of field trials and laboratory investigations, major design changes have been made to improve the system's overall reliability and performance, and to reduce inherent operating inefficiencies. In addition, a survey tool has been incorporated into the drilling tool to determine the trajectory of lateral holes. Further refinements are expected to significantly improve the reliability and performance of the tight radius drilling system and lead towards commercialisation of the technology.

Major advances are also being achieved in the field of mining automation. A key objective of automation research has been to remove workers from hazardous environments, allowing them to operate and manage equipment from a remote location. Automation also promises economic benefits: increased equipment use, reduced maintenance costs and improved product quality.

While there have been many attempts to develop autonomous underground vehicles, the need for complex infrastructure, including some form of beacon system, has been a major drawback. The system of automating underground vehicles (such as haulage trucks) being developed in Australia will be a significant improvement over other systems being developed around the world. In the new system, auto guidance is achieved using typical bandwidth mine radio communications equipment in a module that requires minimal infrastructure and is suitable for fitting to existing remote controlled vehicles. A mapping technique to provide essential information for navigation has also been developed.

Automation research is also focusing on draglines, with a project aiming to increase productivity of this huge piece of mining equipment by up to 4 per cent, a figure which would translate into multi-million dollar savings for a typical

AUSTRALIAN SCIENCE

> **ABOVE:** The PricewaterhouseCoopers research and development investment and innovation group (RDII) can assist organisations at any phase of growth or change in their innovation cycle.

A GUIDING HAND THROUGH THE INNOVATIVE CYCLE

PRICEWATERHOUSECOOPERS (PwC) was recently approached by an Australian company to help launch their innovative micro co-generation technology into the energy distribution market. PwC assisted in a range of activities required for the launch:
> A comprehensive business plan was jointly developed which formed the basis of growth plans and discussions with potential equity providers.
> Financial modelling was undertaken to present a credible revenue growth strategy.
> The presentation was made to potential investors; the company now has a range of investors, increasing its valuation from $5 million to $45 million.
> PwC has been extensively involved in facilitating introductions to large utilities as potential customers and assisting in negotiations with these customers. The company is on the verge of a $1 billion deal with one such customer.
> PwC also provided Government grant and R&D tax concession advice on an ongoing basis to maximise public assistance to the company.
> PwC also gave tax structuring advice which was relevant to its high growth and international strategy.
> The firm is also identifying, quantifying and maximising revenue streams which may be available as a result of a variety of greenhouse/carbon credit audit legislation and regulation throughout the world.

PricewaterhouseCoopers is the world's largest professional services organisation. Drawing on the knowledge and skills of more than 150,000 people in 150 countries, they help clients solve complex business problems and enhance their ability to build value, manage risk and improve performance.

The PricewaterhouseCoopers research and development investment and innovation group (RDII) can assist organisations at any phase of growth or change in their innovation cycle.

The RDII group has experts who can deal with early stage funding and provide structural advice, expertise in government grants and assistance in managerial and strategic matters that impact R&D and innovation decisions.

>> **www.pwcglobal.com/autice**
and turn to page 179 for directory details

MINING AND RESOURCES

Australian coal mine. A dragline control system has been successfully tested under field conditions for the whole operational cycle, from lifting and swinging a full bucket to load dumping and return.

The technical lead in mining and resource science Australia enjoys is set for a significant boost with construction of the $37 million Australian Resources Research Centre in Perth—a strategic alliance between the CSIRO and the West Australian government. This 21st century facility is well placed, in the state that produces two-thirds of Australia's non-fuel minerals and about half its petroleum, and will enhance the international reputation of the CSIRO's Petroleum and Exploration and Mining divisions. Several specialised units from Curtin University will also be based in the Centre, which will encourage collaborative research through CRCs and enjoy backing from major resource companies.

Initial research at the new facility, expected to open in early 2001, will concentrate on technologies that enable the discovery of new, high-quality mineral deposits. New mining methods which cut project development and production costs will also be a priority for scientists.

Two important spin-offs from the Centre's establishment are already occurring—the Centre is attracting scientists with international experience, and there is now a joint CSIRO–universities scholarship scheme for postgraduate research into exploration, extraction and processing of minerals and petroleum.

An important contribution to value-added minerals processing in Australia is being made by the Perth-based AJ Parker CRC for Hydro-

PICTURE: PHOTOLIBRARY.COM

> **ABOVE:** Aerial view of an offshore oil rig

metallurgy. The Centre is one of the leading CRCs, operating in a sector which adds more than $10 billion to the value of minerals produced in Australia each year, and is recognised as one of the most powerful and successful hydrometallurgical R&D groups in the world. The Parker Centre II has considerably more industry commitment than the original centre, with mining majors such as BHP, Comalco, Normandy Mining, Pasminco, Rio Tinto and Western Mining all committing resources. There will be a shift in research focus in the new Centre towards improving process efficiency; another important change will be an increased emphasis on engineering.

With hydrometallurgical processing of nickel laterites at high pressures and temperatures now a commercial reality, the Parker Centre II has also taken the research initiative to improve the mineral industry's understanding of the science and technology involved in pressure leaching.

Several projects are focused on the gold mining industry, with research suggesting that a significant part of the $100 million spent each year on cyanide (as a reagent) could perhaps be saved through better reagent management. Scientists have also teamed with a biotechnology company to work on recycling leach effluents and other waste streams from gold processing. This has considerable environmental benefit and is important to mine operations where clean water is in short supply.

While the mining sector is set to reap the benefits from a wave of new investment in science facilities, the world's longest running mineral processing research project is still winning accolades after 38 years. The so-called P9 Project began in 1962 at a disused mine site in suburban Brisbane. Over the past four decades the project has grown to be the largest collaborative, multifaceted mineral processing research and development project funded by industry anywhere in the world.

P9 began as a collaborative effort between the University of Queensland's Julius Kruttschnitt Mineral Research Centre and the Australian Mineral Industries Research Association (AMIRA). It was the ninth AMIRA project, and hence became known as P9. A range of research projects are conducted under the P9 umbrella, all aimed at optimising mineral processing operations, and all using modelling and simulation to do this.

Research is carried out in collaboration with the University of Cape Town in South Africa and McGill University in Canada, and the project is supported by 38 companies in five countries: Australia, Canada, South Africa, Indonesia and the United States. Because of this long-term support from industry, P9 now has a large number of completed site investigations and thesis studies. These have been consolidated, and have produced significant new methodologies in process design and optimisation. The collection of reliable scientific information over such an extended period has built one of the most comprehensive research databases in the world.

The use of postgraduate students to carry out site-related research has produced more than 80 skilled graduates with Masters and PhD training in Australia alone. These graduates have had a significant influence on the practice of minerals processing worldwide.

Australia is also at the forefront of new metals technology, with a joint CSIRO–industry collaboration paving the way for the production and processing of magnesium. The world's lowest-cost technology for producing high-purity magnesium metal has been developed to tap a large deposit of high-grade magnesite discovered in central Queensland. A 1,500 tonnes a year demonstration plant has been commissioned, with a billion-dollar facility capable of producing 90,000 tonnes a year expected to be operational by 2003. A plant in South Australia—SAMAG—is also close to being built.

Mining and minerals science in Australia has a deserved international reputation for excellence which will only be enhanced by new infrastructure spending. A stable political climate and funding commitments at both federal and state level make Australia an attractive option for mining companies seeking maximum value from their research dollar.

MINING AND RESOURCES

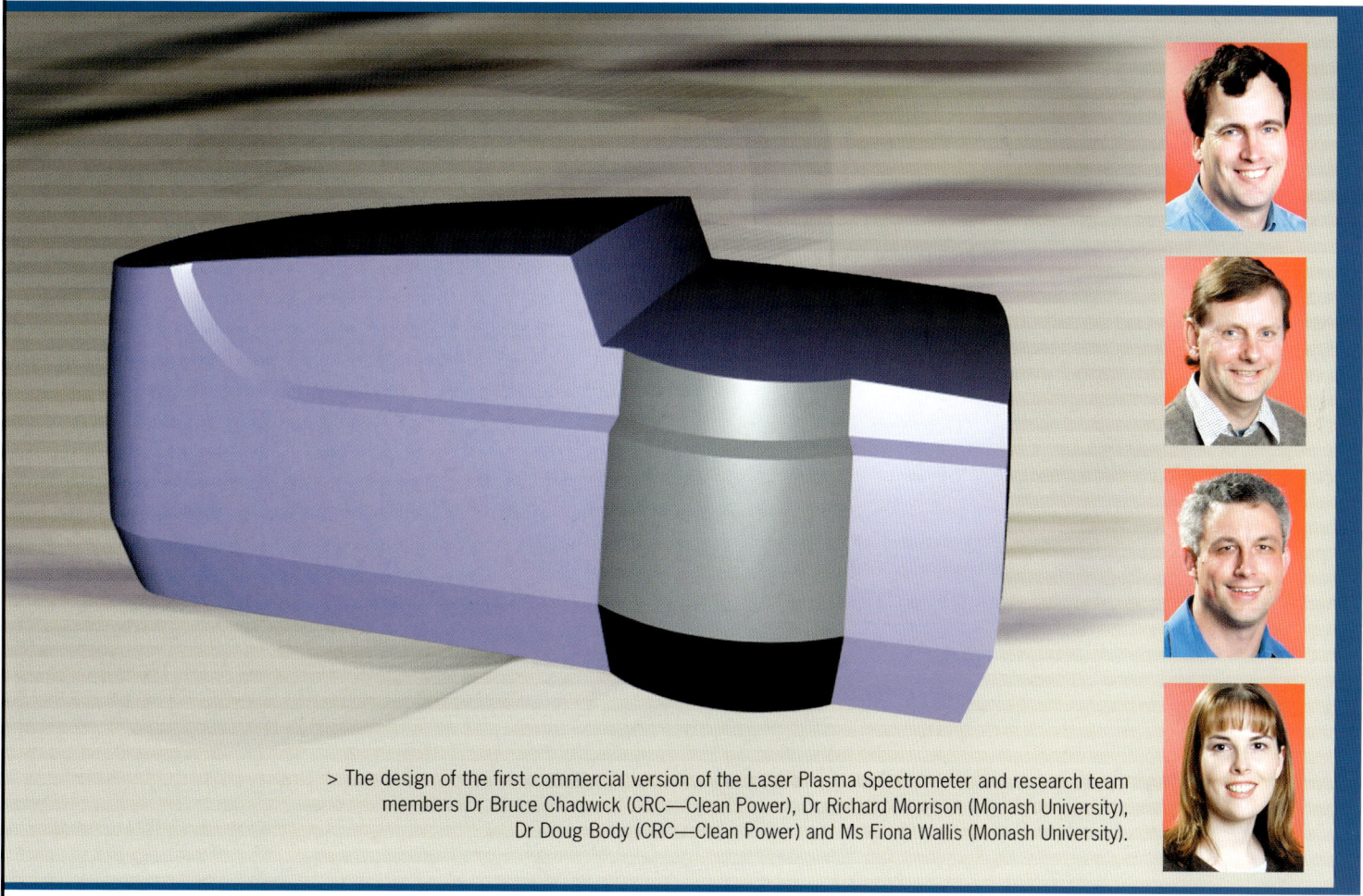

> The design of the first commercial version of the Laser Plasma Spectrometer and research team members Dr Bruce Chadwick (CRC—Clean Power), Dr Richard Morrison (Monash University), Dr Doug Body (CRC—Clean Power) and Ms Fiona Wallis (Monash University).

A NEW TECHNOLOGY FOR MATERIAL ANALYSIS

THE COOPERATIVE RESEARCH CENTRE FOR CLEAN POWER FROM LIGNITE has developed a new Australian instrument which enables rapid determination of the elemental makeup of complex materials, such as minerals, metals and glasses, to detect major components and contaminants.

The laser plasma spectrometer (LPS) has widespread application in industries that rely on maintaining a uniform quality of process materials to ensure the integrity of finished products and trouble-free operation of manufacturing plant. Such industries include minerals processing, building materials manufacture and metal and alloy processing.

The spectrometer was developed to enable the power industry to assess the quality of incoming coal. It addresses a global industry need for better technologies for the characterisation of materials. The technology can be adapted to such diverse applications as the sorting of recycled glass bottles, the monitoring of environmental pollutants and as an educational tool in universities.

The LPS incorporates a high-power laser that induces a bright spark (or plasma) at the surface of a material. The light emitted by this spark is then analysed by a unique spectrometer and detection system. The observed fluorescence for each element is directly related to the concentration of that element in the material being analysed.

The LPS is sensitive to a wide range of elements, including those with low-atomic numbers, such as hydrogen and carbon that are not easily detected by alternative methods. The small amount of preparation required for analysis also enables a high throughput of samples to be maintained.

International patents and proprietary designs protect the LPS technology. It was developed by the CRC team led by Dr Bruce Chadwick, in collaboration with Monash University researchers. Several instruments have been installed in commercial laboratories for further testing and proving, and subsequent commercial design has been undertaken (see photograph). A commercial version of the instrument is being prepared for release in 2002.

>> www.cleanpower.com.au
and turn to page 174 or 184 for directory details

Research at the University of Sydney's Key Centre for Polymer Colloids, led by Professor Bob Gilbert (pictured), is revolutionising polymers, the giant molecules which make up many natural and synthetic materials
> RIGHT: IPRI, in collaboration with Industrial Research Limited (NZ), have developed a prototype electronic nose using conducting polymer sensors (foreground). The electronic nose has been developed for food and beverage analysis

MATERIALS SCIENCE

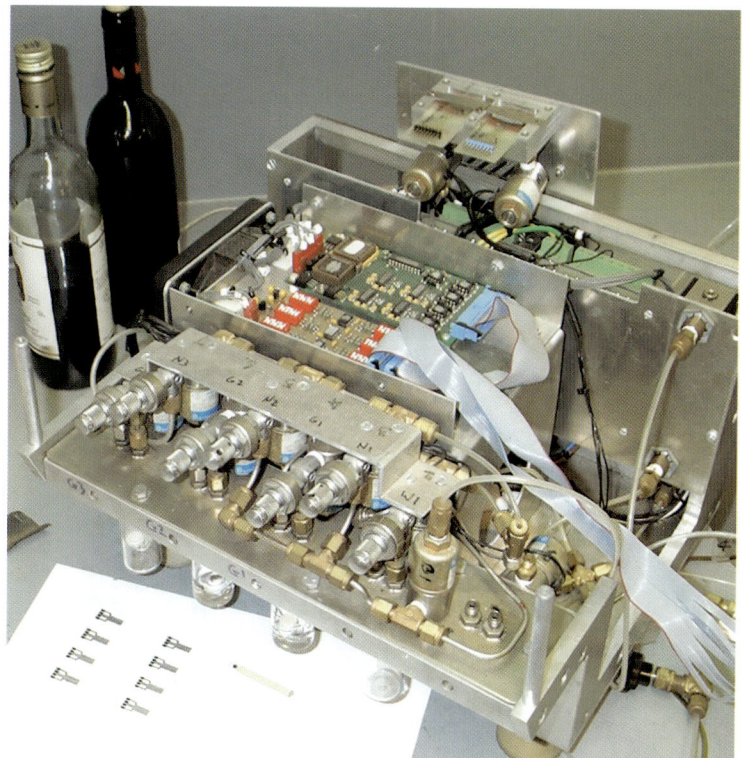

MATERIALS SCIENCE

AUSTRALIAN MATERIALS SCIENTISTS ARE WORKING CLOSELY WITH INDUSTRY, AND IN MANY DISCIPLINES, FROM BIOCHEMICALS AND TEXTILES TO CONSTRUCTION AND ELECTRONICS, REPORTS AMARA BAINS

Materials science research is focused on extending or diversifying the functionality of a variety of materials used in industry. One of the primary ways this is achieved is through the creation of intelligent systems within the materials themselves. These systems enable the materials to sense changes in their environment and make the necessary adjustments to their own properties.

Australian scientists are among the world leaders in materials science research; the application of the developments of this research benefit many industries, including construction, manufacturing and even health.

There are several educational institutions and commercial ventures around the country that work in materials science research, including the Intelligent Polymer Research Institute (IPRI) at the University of Wollongong, the CSIRO, BHP, Visy Board, DuPont and Dionex. The collaboration between academic institutions and industry within Australia and internationally has allowed materials science research to remain a highly productive and commercially viable field.

Biomonitoring fabrics is perhaps one of the most exciting areas of materials science research. Fabrics are modified to act as strain gauges, providing real-time information on strain and pressure encountered

AUSTRALIAN SCIENCE

> **ABOVE:** The new, swaged rib, manufactured from advanced composite materials using the Resin Film Infusion process.

NEW COMPOSITE STRUCTURES BRING SIGNIFICANT WEIGHT AND COST SAVINGS

THE COOPERATIVE RESEARCH CENTRE FOR ADVANCED COMPOSITE STRUCTURES (CRC-ACS) has been developing structural elements for advanced composite aircraft wings since early 1995.

The CRC-ACS has an unequalled reputation for the comprehensive design, analysis, manufacture and testing of lightweight, carbon-fibre reinforced, polymeric ribs.

In recent full-scale tests of a composite wing for a new 100 seat airliner, two developmental ribs performed successfully. The full wing tests included fatigue testing to 90,000 flight cycles and static tests at 'limit' and 'ultimate load' conditions. Through the combined use of advanced composite processing technologies and a new, swaged design, weight savings of up to 20 per cent were demonstrated together with 15 per cent savings in manufacturing costs.

The CRC-ACS is also developing a load introduction rib for the composite outer wing proposed for the new 550-seat Airbus A380 airliner. Innovative test methods have confirmed that swaged stiffeners have superior performance and are lighter and cheaper to manufacture.

>> **www.crc-acs.com.au**
and turn to page 184 for directory details

in exercise and training routines, potentially heralding a new era in health bandages and prosthetics. One example, developed by IPRI, is the 'smart bra', crafted from a new generation of intelligent fabrics. It functions so that the bra will tighten and loosen its straps or stiffen and relax its cups to restrict breast motion, preventing breast pain and sag.

The smart bra is the first of a range of smart textiles projects conducted by IPRI in conjunction with the Biomechanics research laboratory. IPRI is collaborating with the University of Pisa in Italy and the Smartex company to produce new health-monitoring clothing. These biomonitoring fabrics are just one example of the research being conducted by IPRI, which has four other key areas of research: sensors, actuators, energy conversion and energy storage.

In the area of sensors, IPRI has formed an association with Industrial Research Limited, a New Zealand organisation, and the University of Pisa, through which they have developed an electronic nose for olive oil characterisation. The device is being further developed for the detection of micro-organisms, in collaboration with TECRA, a Sydney-based company that develops and manufactures rapid test kits for use in the food, health care and cosmetics industries.

The other areas of IPRI's research—actuators, energy conversion and energy storage—are a significant part of the group's four-year research strategy. In the actuators field, advances have been made in the development of solid polymer electrolytes for use in solid state actuators, and work is currently under way to increase the efficacy of both. An important result of IPRI's work in this field, done again as part of an international collaboration, was the first demonstration of nanotube actuators —a report on this research was published in the prestigious journal *Science* in May 1999.

Energy conversion research set up in collaboration with Massey University in New Zealand has developed the use of novel light harvesting polymers. These show advantages over current systems when used in polymeric photovoltaic devices. Another collaborative approach, this time with CSIRO Molecular Science, has produced promising developments in the production of polymers for hydrogen generation.

In energy storage research, development of a self-powered mobile sensing system as a solution to air monitoring is the prime focus of IPRI efforts. In addition, there is work being done on the development of batteries and supercapacitors based on fabrics.

Some of the recent developments in materials science research have been readily adopted by industry for further development and commercialisation. One such success story is the revolutionary new technology developed by the CSIRO to engineer the surface of polymers and polymeric composites. This technology has been licensed to a US building products company and an Australian car manufacturer, and will provide adhesion strength and durability for both painting and bonding applications that are significantly better than current methods of polymeric substrate pre-treatment. It also brings cost savings by allowing the use of cheaper materials and more efficient processes. The process meets global environmental policies on the elimination of ozone-depleting substances (including solvents and chlorine-based materials) and can be easily integrated into existing manufacturing systems enabling treatment of products, in a production line, at speeds up to 300m/min.

The Building, Construction and Engineering Division (BCE) of the CSIRO has developed a fundamental understanding of how surfaces degrade by combining electrochemistry, surface science and corrosion science. This understanding will be used to promote the development of a new generation of durable structures.

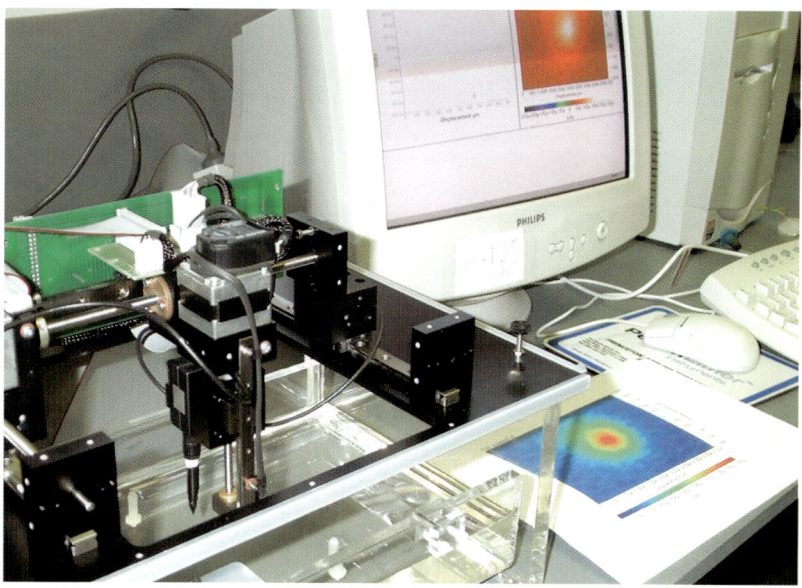

PICTURE: CLINT KINCAID AT IPRI

> **ABOVE:** IPRI has recently established this electrochemical mapping facility. Two complementary instruments enable the mapping of electrochemical events (corrosion etc) on sample surfaces at micron resolution. The facility will initially be used to study corrosion protection of metals using conducting polymer coatings.

SOME OF THE RECENT DEVELOPMENTS IN MATERIALS SCIENCE RESEARCH HAVE BEEN READILY ADOPTED BY INDUSTRY FOR FURTHER DEVELOPMENT AND COMMERCIALISATION

AUSTRALIAN SCIENCE

The BCE's accelerated weathering chambers and environment analysis allow comparison of characteristics across similar products, thus permitting identification of materials that are best suited to manufacturers' requirements. The analytical capabilities of the CSIRO are supported by their extensive knowledge of the service environments within Australia, plus their access to corrosivity databases—all this makes sure their results are highly regarded by regulators and standards bodies.

The importance of materials science research is often highlighted in the construction industry.

carbon graphene sheet is 'rolled-up' to form the nanotube. Discovered about 10 years ago, a recent display of their capacities has highlighted the potential of this material. Nanotubes with an elastic modulus (stiffness) of around 1000 GPa (compared with steel, at 200 GPa) also have high strength and resilience and provide exceptional mechanical properties, making them exciting new building blocks for nanodevices—miniaturised electronic devices smaller than microchips. However, the application of this technology is not limited to nanodevices; it also provides

THE APPLICATION OF DEVELOPMENTS IN MATERIALS SCIENCE RESEARCH CAN MEAN WORKING ACROSS MANY OTHER TRADITIONAL DISCIPLINES—FOR INSTANCE BIOMECHANICS, TEXTILES AND ELECTRONICS

For example, pipeline durability and plastics performance is dependent on the materials, the installation, the service parameters and the environmental conditions under which the pipes and plastics are used. Australian research is developing early detection procedures to establish the degradation rates and failure mechanisms of pipeline materials under a range of conditions. This research will predict, establish and then extend the lifetimes of these materials, and the knowledge gained from this research will be used to develop new material applications and material additives.

On a much smaller scale, one of the most promising areas of innovation in materials science is nanotubes. Nanotubes are highly conductive hollow tubes of carbon that can behave as semiconductors, depending on the way in which the

extremely interesting opportunities for micro and macro applications. Some of the new and exciting areas that nanotube technology can be applied to are the production of artificial muscles and energy storage.

The application of developments in materials science research can be difficult, as there is a need to work across many other traditional disciplines—in the case of the smart bra, for instance, a collaboration between biomechanics, textiles and electronics was required to produce a prototype.

While there is some support from local industry, a number of institutions receive considerable overseas backing. Notwithstanding funding-related issues, Australia remains on the cutting edge of new developments in materials science and provides world-class educational opportunities for local and international scientists.

MATERIAL SCIENCE

PLASTICS SIMULATION SOFTWARE WINS CUSTOMERS WORLDWIDE

THE COOPERATIVE RESEARCH CENTRE FOR INTELLIGENT MANUFACTURING SYSTEMS AND TECHNOLOGIES (CRCIMST) runs a number of research programs focusing on the Australian machine tool industry, plastics processing and machining, aerospace design and assembly, metal cutting and forming and new generation machine specification, design and control.

One Australian company whose success is due in part to the CRCIMST is Moldflow International. The company's founder, Colin Austin, first developed plastics injection moulding software as a university research project. He then created his own company to develop and market the product.

When Moldflow International became a founding member of the CRCIMST in 1993, the CRC program enabled the company to establish a research project far more ambitious than any it had previously attempted.

Together, they gathered together a team of engineers and scientists from CRCIMST partners in Australia and from around the world. Their task was to investigate the process called 'multiphase injection moulding'.

Early on in the project, however, the opportunity arose to develop a new 3D engine which had the potential to revolutionise design in the plastics industry. They had discovered a way to develop energy and material savings for plastics manufacturers and the possibility of reusing recycled polymers.

The CRC program demonstrated its flexibility and ability to react to industry needs by changing direction to take advantage of this new opportunity. The result was a product called 'Flow3D' — a plastics simulation software program which solves issues previously not addressed by techniques used to analyse traditional, thin-wall plastic part and mould designs.

The development of this new software confirmed Moldflow International's position as the world's leading innovator in the field. Customers already taking advantage of the Flow3D solution include 3M Company, BASF, Corning Precision Lens, DaimlerChrysler, GE Plastics, Hitachi, Honeywell, Mitsubishi Rayon, Molex, Mont Blanc-Simplo, Robert Bosch, Samsung Electronics and Valeo Electronique, as well as more than 25 major universities worldwide.

>> **www.crcimst.com.au**
and turn to page 185 for directory details

Communications tower
> RIGHT: Personal Digital Assistants are becoming more and more popular as they become able to be used for more, and more complex, tasks

TELECOMMUNICATIONS

TELECOMMUNICATIONS

TELECOMMUNICATIONS IS TURNING THE IMPOSSIBLE INTO THE EVERYDAY, IN AREAS AS DIVERSE AS HEALTH CARE COMMUNICATIONS, TELETRAFFIC AND MINE COMMUNICATIONS, REPORTS AMARA BAINS

Global statistics on mobile phone use, the number of people connecting to the Internet and the billions of e-mail sent around the world daily make a compelling case for the argument that the telecommunications revolution has landed—permanently.

Telecommunications research has produced results that have had applications ranging from simple domestic use and modern medicine to satellite technology and space exploration. There are many institutions and collaborations, both academic and commercial, exploring the realms of telecommunications science, with much of the research seeking to improve existing services and meet the growing demands and needs of the community at large.

At the University of New South Wales, a collaboration between medical and engineering departments and partnerships with the private and public health sectors, government and other academic institutions have led to the establishment of the Centre for Health Informatics (CHI). This centre engages in research, development and commercialisation of advanced information and communications technologies for health care delivery.

It has four key areas of research, all of which seek to improve health care by providing technological

AUSTRALIAN SCIENCE

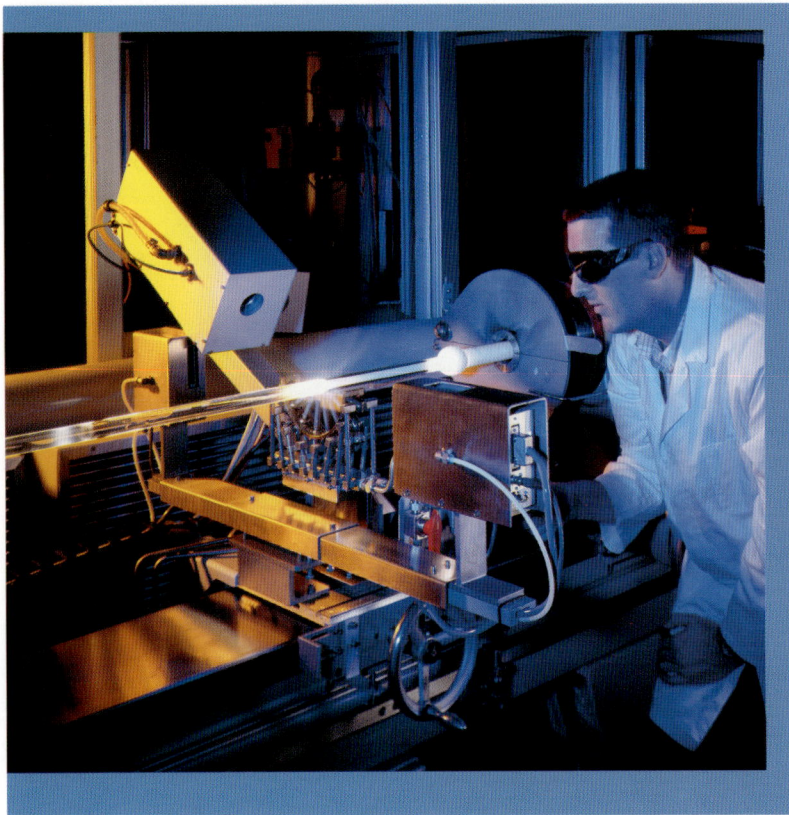

> **ABOVE:** Production of special optical fibre at The Australian Technology Park in Redfern, Sydney.

INFORMATION STREAMS COMBINED WITHIN AN OPTICAL NETWORK

A subsidiary of the REDFERN PHOTONICS GROUP has recently released the RBN GigaWave™, one the latest products to be developed out of research generated by the Australian Photonics Co-operative Research Centre (CRC).

Produced by Redfern Broadband Networks, the RBN GigaWave™ system combines a variety of information streams such as video, telephone and data on single fibres by giving each stream its own wavelength. Using RBN GigaWave™ makes the total capacity of each individual fibre 100,000 times greater than existing ethernet local area networks and enables the management and control of all data and wavelengths within an optical network.

The Australian Photonics CRC is a joint venture between 6 research and 24 industry participants. It is equipped with state-of-the-art facilities and involved in research and development, commercialisation, education and training. Australia currently has 1.2 per cent of the global market for photonic products, a figure which is growing at more than 20 per cent per year. Redfern Photonics and the Australian Photonics CRC are helping grow Australia's market share in this dynamic industry.

>> www.redfernphotonics.com
and turn to page 185 for directory details

solutions for health care professionals and consumers. The first, the evidence-based decision support program, aims to develop ways of providing online access to clinically relevant information that will help clinicians and consumers make better decisions.

Clinical communications is an area concerned with both understanding the importance of communication in health care delivery and experimenting with new technologies. One example of this experimenting is the WAP-enabled shared task list for doctors in hospitals.

The wireless application protocol (WAP) is a means of accessing the World Wide Web from a mobile phone. Applying this technology in hospitals, particularly for staff on the move, such as doctors, will allow them access to information such as patient records, drug databases and staff directories, will help them organise activities collectively, keep track of each other's activities and place requests for tasks they need carried out.

Another interesting area of research by the CHI is the home telecare project. As the name suggests, this project involves research into communication technologies that would allow patients to be monitored in their homes. Information from this home monitoring would be passed on to the patient's primary care giver—the aim is to improve clinical outcomes, and allow the elderly and the chronically ill to stay at home longer.

The home telecare project includes developing a 'smart home' system that unobtrusively monitors the activity level and health status of elderly people. Its advantage is its ability to detect deterioration in health much earlier than current alert systems, which are usually triggered by crises such as falls. Further potential lies in ambulatory monitoring of physiological parameters such as ECG and respiratory function, and linking these monitoring devices to the World Wide Web so that a remote clinician can look in to check the health status of chronically ill patients.

The fourth major research area of the CHI is in evaluation—assessing the effectiveness of new information and communication technologies in improving health outcomes and delivery.

The use of telecommunications in health is a burgeoning area, and many companies normally not associated with health are joining forces to

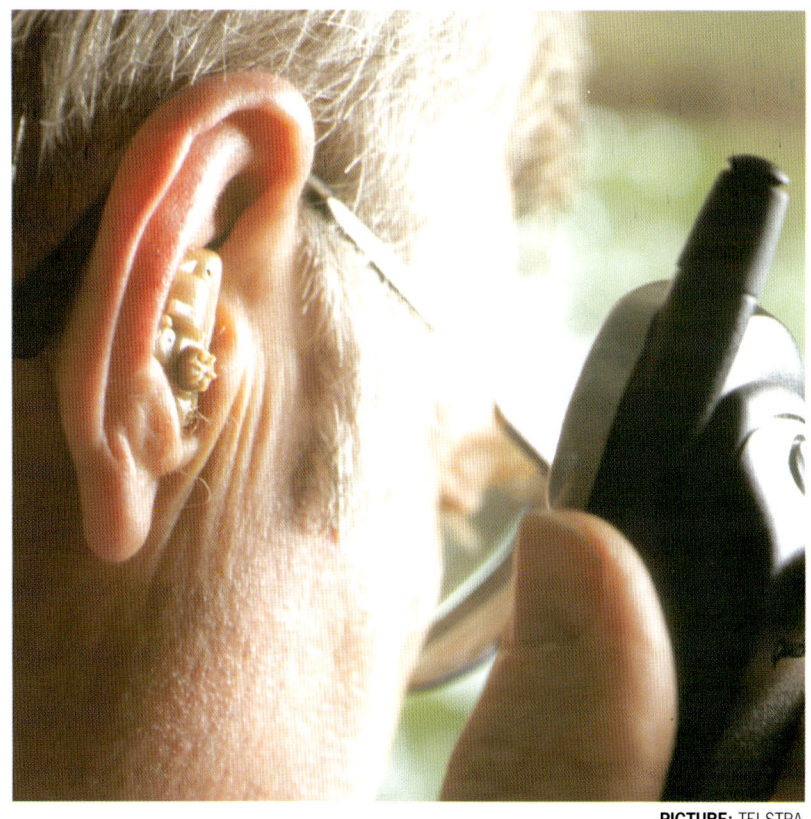

> **ABOVE:** Telstra is finding ways to allow the hearing impaired access to mobile phones

THE AIM OF THE HOME TELECARE PROJECT IS TO IMPROVE CLINICAL OUTCOMES, AND ALLOW THE ELDERLY AND THE CHRONICALLY ILL TO STAY AT HOME LONGER

produce mutually beneficial outcomes. Australian telecommunications giant Telstra, for instance, has established a relationship between its research arm and the Bionic Ear Institute to develop models for the way in which the brain processes the speech signals it receives from the ear. The models can be used in a variety of applications, such as improving the performance of the cochlear implants for the deaf and improving speech recognition accuracy.

Interactive Voice Response (IVR) platforms are behind telephone services—such as telephone banking and bill payment—which require the user to press buttons in response to pre-recorded instructions. New technology would mean that users of the service could simply speak their instructions into the phone, have them interpreted by the computer and then be connected to the appropriate service without having to listen to a menu.

Telstra Research is involved with a number of other projects and collaborations. In work on third generation mobile phones Telstra is involved with Lucent Technologies in the United States and the United Kingdom, trying to better understand what customers want from third generation mobile phones—what services will be relevant to their lives and at what price.

In an initiative to encourage e-business, Telstra has developed digital certificates, which it issues to people and computer systems to ensure online security. To develop this concept for global markets, Telstra has licensed the software to an Australian high-tech company, Adacel.

Telstra Research also continues to work in the areas of software engineering, artificial intelligence and neural networking, systems integration, interface design, broadband services and optical fibre technologies.

Some of the highlights of Telstra's research include the Application Development Automatic Generation Environment (ADAGE), a tool that will allow businesses to automatically implement software changes without relying on sophisticated IT skills or coding experience, and Condor, which provides call traffic managers with real-time visual representations of call volumes. Condor will allow effective management of mass calling events and protection of customers from the effects of an overload situation.

A recent development preparing for commercialisation is Internet call waiting. It provides a virtual second line, enabling a subscriber with a single telephone line to access the Internet without missing incoming phone calls.

Telstra works with a number of academic institutions in its research programs. The Telecommunications Information and Technology Centre (TITC) at the University of Wollongong, and specifically its Switched Network Research Centre, was recognised as a centre of excellence by Telstra and has received significant funding from Telstra and Motorola for research into telecommunications, the Internet, speech and multimedia. It also carries out funded research for Ericsson, AWA and the Defence Services Technology Organisation.

The TITC was established to investigate the design and performance of advanced telecommunications network technologies, and its activities include network design for new network technologies and architectures. Examples of the TITC's developments include Asynchronous Transfer Mode (ATM), the new standard for high-speed data transmission, which works by sending small packets of information asynchronously, and xDSL, where x can refer to Asymmetric or Very high speed digital subscriber line (ADSL or VDSL, respectively), both designed to increase the speed of transmission of data.

A number of other Australian universities also have leading centres in telecommunications research. At RMIT University in Melbourne, the Centre for Advanced Technology in Telecommunications (CATT) was set up to focus on research into teletraffic engineering, telecommunications software tools, telecommunication network modelling, digital signal processing and integrated optics. It also works in other research areas, such as computer-aided software engineering techniques, microwave power engineering antennas, applied electromagnetics and mobile and personal communications.

> THE **UNIVERSITY CENTRES** ARE OFTEN STRATEGIC PARTNERS IN DEVELOPING TECHNOLOGY, AS WELL AS PROVIDING THE SKILLED INDIVIDUALS TO WORK IN COMMERCIAL RESEARCH VENTURES

TELECOMMUNICATIONS

PICTURE: TELSTRA

> **ABOVE:** Condor providing real-time visual representations of call volumes for traffic managers

The university centres are often strategic partners in developing technology, as well as providing the skilled individuals to work in commercial research ventures.

One such venture is the Motorola Australian Research Centre (MARC). Created in 1995, MARC focuses its research on technologies for speech, image and video processing, networks and communications systems. It develops algorithms and proof-of-concept implementations in speech recognition, speech coding, communications and networking, image coding and video coding.

The development work of the Communication and Network Lab, one group within MARC, focuses on connection solutions for devices that interconnect via the Internet. Divided into two groups, the Communications Systems and Technology group works on physical and link-layer technologies, with an emphasis on wireless and cable networking, while the Networking group works on the network layer, with the emphasis on the scalable and ubiquitous deployment of Internet-aware devices.

The visual image processing lab develops multimedia technologies for Motorola's mobile and Internet and radio terminals, and the group researching speech technology looks at developing algorithms that enable real-world human–machine communications via voice. Using natural language and understanding to enable transparent user interfaces between the human and machine world is a prime focus of research in the lab.

As with most areas of Australian scientific research, the CSIRO also has many divisions

AUSTRALIAN SCIENCE

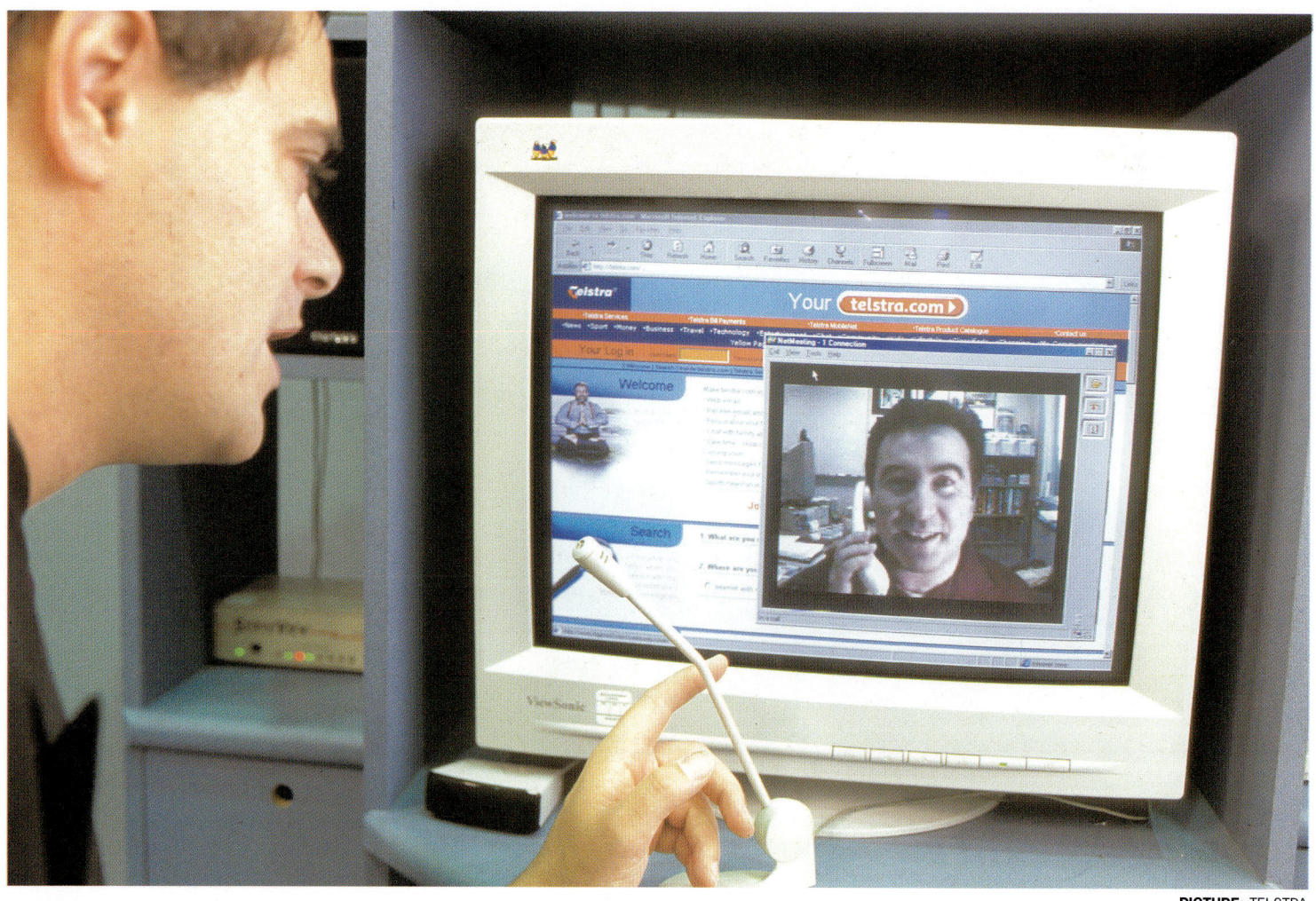

> **ABOVE:** Telstra's Intelligent Network

PICTURE: TELSTRA

working in the field of telecommunications. Its Information Technology and Telecommunications Division has extensive capabilities in spatial information systems and technologies for electronic documents, and is focusing its research programs on the areas of broadband wireless systems, networking for the information economy, and microwave systems for strategic self-reliance— the latter is worked on in conjunction with the Department of Defence, and concerns the development of Australian technology in areas where overseas suppliers may be unwilling or unable (by government regulation) to supply the latest equipment and information, as was the case in the Gulf War, for instance. Some of the research results of this Division include a two-way radio for underground mines (surface to miner and return) which will act as an emergency communications tool permitting swift action in the event of a disaster.

The Centre for Telecommunications and Industrial Physics (CTIP) at the CSIRO also has many ongoing projects and collaborations, with one recent success being the design of a revolutionary new multibeam antenna which replaces many other systems, thus requiring fewer antennas and reducing associated running costs of numerous antennas—this has just been acquired by a German

TELECOMMUNICATIONS

company, TST Kommunikations-technik GMbH, for use in the lucrative European pay TV market. To assist industries such health, retail, online services, finance, banking security and defence, the CTIP is also channelling its expertise into image acquisition, interpretation, transmission and storage. The research programs of the CTIP often involve establishing linkages with multinationals and their Australian alliance partners.

The work of the CTIP covers many areas. Broadband networked telecommunications research is another area currently being tackled

CSIRO became a principal member, and it has contributed to the standardisation process.

High-speed radio modems are being used in a number of applications, from microwave links to the distribution of data over cable networks and wireless Internet access, and the CSIRO has developed a high-speed digital modem testbed to facilitate rapid response to customer requests for modem designs at a variety of data rates and modulation formats.

Applications of the research carried out by the CSIRO are finding their way into many services not

THE CSIRO INFORMATION TECHNOLOGY AND TELECOMMUNICATIONS DIVISION IS FOCUSING ITS RESEARCH PROGRAMS ON BROADBAND WIRELESS SYSTEMS, NETWORKING FOR THE INFORMATION ECONOMY, AND MICROWAVE SYSTEMS FOR STRATEGIC SELF-RELIANCE

by the CTIP. This area focuses on the design and management of networks as engineering systems that will support applications requiring negotiable Quality of Service. ATM technology, with its explicit Quality of Service features, provides the initial focus, but the scope has been expanded to include other concepts.

Wireless ATM research has grown from wireless Large Area Networks (LANs), and will provide an important role in broadband communication networks of the future. Collaboration with Macquarie University developed a Medium Access Protocol ('parrot MAC') which is suitable for carrying wireless ATM traffic—the future direction of telecommunications. When the United States-based ATM forum started a Wireless ATM Group, the

necessarily associated with telecommunications. The development of a low-cost high-precision location system which provides superior accuracy, resistance to interference and better reliability than the Global Positioning System (GPS) will allow compact transceivers in multiple mobile units to be tracked in real time to within millimetres. This system is being tested in the tracking and automated control of mining vehicles and farm machinery, the tracking and arrival time estimation of public transport and in inventory location (for containers, and precious or dangerous goods).

The CSIRO, like many other research organisations, forms cooperatives with other organisations to share expertise and facilitate research and development. One group the CSIRO participates in is the Australian Telecommunications Cooperative Research Centre (CRC), which specialises in four broad research areas: application programs, networking programs, and wireless and enabling technology programs.

The application programs consist of next generation Internet research, advanced video retrieval services and multimedia over wireless networks. The programs trial these applications, which are thought to be representations of what users will be able to expect from future networks. The trialling will expose Quality of Service issues and point to new types of network services that may be required.

Handling real-time traffic on packet networks is the main objective of the network program. The three projects underway are ATM Adaptive layer 2 (AAL2) traffic management and switching, which allows the packets of information sent by ATM to be converted back into information that can be used by the receiving device, real-time signal transfer over Internet Protocol (IP), and real-time traffic measurement.

Wireless program research is developing telecommunications services that allow users to remain 'untethered', that is, able to use services anywhere, anytime. Applications range from the traditional voice services over real-time video to high-rate data and multimedia services. The projects under this program include development of a wireless Code Division Multiple Access (CDMA) scanner, smart antennas, which will be used to minimise problems arising from the increased volume of teletraffic over the mobile communications networks, software-defined radio, which sets up the flexible radio platforms necessary to support mobile service diversity, coding and modulation developments to improve signal quality in non-stationary wireless channels, and multiuser detection in wireless CDMA, a technique that can be used to improve capacity and coverage of a CDMA system such as wideband-CDMA.

The enabling technology program of the Australian Telecommunications CRC is biased towards commercially significant technologies. The Centre has demonstrated its expertise in enabling technology through development of many devices that have been adopted by industry, such as a charge coupled device (CCD) camera system for Boeing, low-noise pre-amplifiers for Specterra, a West Australian airborne remote sensing company, and design reviews for Ampcontrol, an electrical and electronics design and manufacturing company based in Newcastle, New South Wales. The two main current research projects in the Centre's program are in electromagnetic capability and signals in electronic and communications systems. Among the core partners of the Centre are Ericsson Australia, Curtin University of Technology (Western Australia), Radio Frequency Systems and Vodafone Network.

Australian telecommunications research organisations have developed many technologies that have widespread application in the community, and the success of these developments and the growth of this sector can be attributed, in part, to the strategic relationships between industry bodies, academic and research groups.

TELECOMMUNICATIONS

> **ABOVE:** Staff from the Centre for Quantum Computer Technology in the Semiconductor Nanofabrication Facility: (left to right) Dr Nancy Lumpkin, Dr Alex Hamilton, Professor Bob Clark (Centre Director), Dr Michelle Simmons, Dr Andrew Dzurak.

QUANTUM COMPUTING—THE QUBIT TAKES CENTRE STAGE

The Australian Research Council has established the Centre for Quantum Computer Technology in an effort to push computer technology further and develop a whole new basis for computer architecture.

Based at the University of New South Wales (UNSW), and led by Professor Bob Clark, the Centre is establishing a research program aimed at producing the fundamental building block of a quantum computer—the quantum bit, or 'qubit'.

Computer technology has revolutionised the world in recent times, and will continue to do so well beyond the year 2000. Operations possible only on the most powerful supercomputers twenty years ago are now routinely performed on a standard PC.

But some of the technological forces that have made this progress possible are reaching their limits. Transistor speed cannot be increased indefinitely, and fundamental limits on miniaturisation will be approached in less than twenty years.

The potential power of quantum computing has encouraged many researchers to pursue the ultimate goal of building an operational quantum computer. But Professor Clark's team has already attracted international attention owing to their unique approach to the problem.

While a few qubit devices have already been constructed using different techniques, none of these methods readily lends itself to large-scale applications. The Centre for Quantum Computing Technology—in partnership with the University of Queensland and the University of Melbourne, together with the prestigious Los Alamos National Laboratory (USA) and the California Institute of Technology (USA)—is investigating the construction of the first solid-state qubits utilising existing silicon-based semiconductor fabrication technology.

The benefits of building quantum computers with a silicon base would be immense. Not only is the current computer manufacturing industry already equipped for mass production of silicon-based components, but it's likely that silicon-based qubits could be more easily assembled into larger-sized quantum computers.

For some applications, a 30-qubit quantum computer could outperform the most powerful conventional supercomputers in the world.

>> **www.arc.gov.au**
and turn to page 172 for directory details

Two antennas of CSIRO's Australia Telescope Compact Array
> RIGHT: Parkes Radio Telescope

ASTRONOMY

ASTRONOMY

AUSTRALIA HAS A DISTINGUISHED HISTORY, AND A BRIGHT FUTURE, IN BOTH RADIO AND OPTICAL ASTRONOMY, AND IN RESEARCH AND COMMERCIALISATION, REPORTS JOHN O'BYRNE

Australia's place in modern astronomy has been shaped by history and geography. The European settlement of Australia followed from the voyage of James Cook to Tahiti to observe a transit of Venus across the sun in 1769. A century later, Australian John Tebbutt was the best known of a small group of astronomers exploring the southern sky, including the unique view it presents of the centre of our galaxy, the Milky Way, and the nearest external galaxies, the Magellanic Clouds.

A further hundred years on, the southern sky presented astronomers with probably the most important astronomical event of recent times—supernova 1987A (SN1987A) in the Large Magellanic Cloud. Since Tebbutt, Australian astronomy had made good use of its southerly vantage point and become a major astronomical centre, well equipped to observe the spectacle of a star ending its life in a supernova explosion.

The largest optical telescope in Australia was then, and remains, the 3.9 metre diameter Anglo-Australian Telescope (AAT) near Coonabarabran, in New South Wales. Together with its companion, the 1.2 metre UK Schmidt Telescope, it is perhaps best known to the public for the amazing astronomical colour photographs taken by David Malin.

The strength of the AAT has always been its people—the astronomers and engineers. In recent years, staff of the observatory have pioneered the use of optical fibres in astronomy, culminating in the Two Degree Field (2dF) instrument, which combines innovative use of large optical components, fibre optics and robotics to turn the AAT into a powerful spectroscopic survey telescope. The 2dF is currently undertaking a survey of 250,000 galaxies and 30,000 quasars, with aims that include determining how galaxies and quasars evolve with time and understanding the scales on which they build clusters. A similar 6dF instrument will soon be in place at the UK Schmidt Telescope.

Telescope (AT). The six 22-metre dishes at the Paul Wild Observatory near Narrabri, New South Wales are used to generate maps of radio sources with 100 times better resolution than the Parkes dish alone. Joined with Parkes and other telescopes across Australia and overseas, they can form a radio telescope thousands of kilometres across, using a technique known as Very Long Baseline Interferometry (VLBI). Among the AT's observing programs is the mapping of hydrogen gas in the Large Magellanic Cloud in unprecedented resolution and detail; this has

JOINED WITH PARKES AND OTHER TELESCOPES ACROSS AUSTRALIA AND OVERSEAS, THE SIX 22-METRE DISHES AT THE PAUL WILD OBSERVATORY CAN FORM A RADIO TELESCOPE THOUSANDS OF KILOMETRES ACROSS

Perhaps even better known to the public than the AAT—especially since the movie *The Dish*—is the CSIRO's 64 metre radio telescope at Parkes, New South Wales. It began operating in 1961, building on the leading place in radio astronomy that Australian researchers had earned after World War II. Parkes continues to work today, with innovative developments such as the multibeam upgrade continuing to improve its performance. Its large size makes it an ideal instrument to track the faint radio whispers of spacecraft voyaging across the solar system, and a useful adjunct to the NASA Deep Space Network dish at Tidbinbilla, in the Australian Capital Territory. In return, Tidbinbilla's ear is occasionally tuned to astronomical projects.

Today, however, the leading edge of much Australian radio work is marked by the Australia

revealed the Large Magellanic Cloud to be punctuated by 'superbubbles', carved by supernovae and stellar winds.

Optical astronomers have long wanted to emulate their radio cousins and link telescopes in order to improve their ability to resolve fine detail. The shorter wavelength of light, compared with radio waves, makes this 'optical interferometry' much more challenging. In the 1960s, the Narrabri Stellar Intensity Interferometer precisely measured the diameters of 32 stars, thereby establishing our understanding of the temperature scale for stars hotter than the sun. This work continues with the Sydney University Stellar Interferometer (SUSI), a world leader in developing the techniques of modern optical interferometry and the largest instrument of its type in the world.

> **ABOVE:** Five antennas of CSIRO's Australia Telescope Compact Array

PICTURE: J MASTERSON © CSIRO

Australian astronomy extends well beyond these instruments, though, with various university groups having earned international reputations for their instruments and research. The Australian National University's 2.3 metre telescope, for example, pioneered some of the modern features of lightweight telescope and enclosure construction. Rather different 'telescopes' have studied showers of sub-atomic particles caused by the highest energy cosmic rays hitting the Earth's atmosphere.

Australian astronomers and their international colleagues continue to do important science with these and other instruments, complemented by the work of the theoreticians, who provide a solid basis to any observational program. SN1987A, for example, remains a subject of intense interest for telescopes in the south and theoretical studies worldwide. After its initial outburst it dropped out of view of radio telescopes until 1991, when it reappeared faintly, first to the Molonglo Observatory Synthesis Telescope (MOST) near Canberra and then to the AT. Its brightness is now building towards the expected collision of the supernova shock wave with the shell of material ejected by the star before it exploded.

Among many other current projects is the Anglo-Australian Planet Search, which is the southern hemisphere component of an international effort to discover planets around other stars. To date, some 70 Jupiter-like planets have

> **ABOVE:** Sydney University's School of Physics, through the Gemini project, is undertaking a major survey of the southern sky which is expected to result in the mapping of more than 500,000 radio galaxies in the Southern Hemisphere

been discovered internationally, using radial velocity measurement. Finding and studying these objects will be crucial in building our understanding of the mechanisms of star versus planet formation.

Astronomy is now dominated by these 'big science' projects, which call for international collaborations and in which Australia plays an integral part. The study of cosmic rays by astronomers at the University of Adelaide is a case in point—it finds them paired with the Japanese in the Cangaroo telescope at Woomera, in South Australia, and part of an international collaboration to build the Pierre Auger Observatory in Argentina.

Australia also has a 5 per cent share of a seven-nation international collaboration in the Gemini project. Optical and infrared astronomy is now the domain of 8–10 metre telescopes, and there will soon be two operational 8 metre Gemini telescopes, one in Hawaii and the other in Chile. Gemini's focus is on achieving super-sharp images and high sensitivity using near infrared light. A small proportion of Gemini time will be reserved for Australian astronomers.

The AAT's important position in this era of telescopes twice its size is marked out by 2dF, which signals a partial shift in focus to wide field survey programs, and illustrates the AAT's growing reputation for world-class instrumentation. One result of this is a contract to build a fibre positioner (affectionately known as the OzPoz) for the Flames spectrograph at the European Southern Observatory (ESO) Very Large Telescope (VLT). Other innovative concepts have been proposed in design studies for instruments on the 8.3 metre Subaru and the 4.2 metre Soar telescopes.

The Australian National University's Research School of Astronomy and Astrophysics (RSAA) is developing the Near-infrared Integral-Field Spectrograph (NIFS) for the Gemini project.

Another product of astronomical research is the Canberra-based company Auspace, a significant player in Australia's space industry. It grew out of work on the Starlab project in the 1980s at Mt Stromlo Observatory near Canberra, and now makes hardware for Earth-observation satellites and telecommunications. Auspace built the Endeavour ultraviolet camera which flew on the Space Shuttle in 1992 and 1995. This was the first major all-Australian space payload to be built since WRESAT (the Weapons Research Establishment Satellite) was constructed in the 1960s.

AUSTRALIAN ASTRONOMY RANKS QUITE HIGHLY IN ITS IMPACT ON THE WORLD SCENE —IT IS WIDELY ACKNOWLEDGED AS ONE OF OUR SCIENTIFIC STRENGTHS

By combining the superb image quality of the Gemini telescope with the high resolution of NIFS, astronomers will be able to study the infrared structure of astronomical objects on scales comparable to those achieved with the Hubble Space Telescope.

Radio astronomers face a less immediate challenge to their dominance of the southern sky than do their optical colleagues. For now, the AT remains the only radio array surveying the deep south. It has recently been extended—by the addition of a new 'arm'—to further improve its imaging performance, and will soon expand its range from centimetre wavelengths to millimetre wavelengths.

The expertise that makes the AT a world leader in its field carries over to commercial technology. Satellite dishes and their receivers use the same technology, and both have benefited from developments where astronomers have led the way.

Auspace will also build the MONS telescope (Measuring Oscillations in Nearby Stars), to be launched aboard the Danish satellite Rømer in late 2003. This international project, also involving the University of Sydney, will build on current work in stellar seismology to determine the internal structures of stars by detecting their pulsations.

Another current Australian project is Federation Satellite One (FedSat-1), a low-cost microsatellite built to undertake communications, space science, remote sensing and engineering experiments. FedSat will be built and operated by the group of research organisations, companies and universities that makes up the Australian Cooperative Research Centre (CRC) for Satellite Systems.

Astronomy in Australia ranks quite highly in its impact on the world scene, despite our having fewer than 200 professional astronomers and lower

AUSTRALIAN SCIENCE

> **ABOVE LEFT:** The Large Magellanic Cloud (LMC)—a satellite galaxy of the Milky Way, 180,000 light-years distant > **ABOVE RIGHT:** A colour composite image of supernova remnant 1E 0101.2-7219—the remains of an exploded star

PICTURE: CSIRO AUSTRALIA TELESCOPE NATIONAL FACILITY

PICTURE: X-RAY (NASA/CXC/SAO); OPTICAL (NASA/HST); RADIO (CSIRO, AUSTRALIA TELESCOPE NATIONAL FACILITY)

levels of per capita funding than comparable countries. Astronomy is widely acknowledged as one of Australia's scientific strengths.

To maintain Australia's enviable position in the astronomical world it is essential to have a strong community of theoretical and observational astronomers with access to the most advanced facilities. Besides membership of Gemini there are proposals to join the European Southern Observatory (ESO), to gain access to their optical and radio facilities in Chile or to seek a collaboration to build an Australian Large Telescope (ALT) there.

Another option is to once again exploit Australia's historical and geographical advantages, this time by developing astronomy in Antarctica. The Joint Australian Centre for Astrophysical Research in Antarctica (JACARA) has been conducting tests at the South Pole to better understand the possibilities of 'astronomy on ice'. Antarctica appears to be the best ground-based site in the world for infrared observations, and a proposal is emerging to build the 2 metre Douglas Mawson Telescope (DMT), with an infrared performance exceeding that of the largest telescopes elsewhere in the world, there.

Radio astronomers have perhaps an even grander vision. An international consortium that includes the CSIRO is studying the technology necessary to build a huge new radio telescope, the Square Kilometre Array (SKA). The SKA will have a collecting area almost 100 times larger than what is now the biggest radio telescope, allowing it to gather crucial new information on the formation and early history of stars, galaxies and quasars, and to play a major role in the search for extraterrestrial intelligence. Given the need for a 'radio-quiet' location, Australia is a likely location for the SKA.

These proposals present exciting opportunities for Australian astronomy. Given sufficient funding to support both people and instrumentation, Australia will be able to retain its distinguished place in astronomy into the early part of the 21st century.

Parkes Radio Telescope

AUSTRALIAN SCIENCE

PICTURE: CRC FOR MOLECULAR PLANT BREEDING

COOPERATIVE RESEARCH CENTRES

Over the past decade, Cooperative Research Centres (CRCs) have proved to be an efficient mechanism for changing research culture and now play a key role in encouraging and assisting industry to update its technological skills in order to harden its competitive edge. The following list includes centres that are playing a critical role in the manufacturing industry. A complete list of CRCs is available from the CRC Secretariat.

CRC SECRETARIAT, Department of Industry, Science and Resources
GPO Box 9839, Canberra ACT 2601
Phone. **02 6213 6429**   Fax. **02 6213 6422**
crc@isr.gov.au
www.isr.gov.au/crc/index.html

COOPERATIVE RESEARCH CENTRES

> MANUFACTURING TECHNOLOGY

CRC for Advanced Composite Structures www.crc-acs.com.au

CRC for Welded Structures www.crcws.com.au

CRC for Polymers www.crcp.com.au

CRC for Bioproducts www.botany.unimelb.edu.au/labs/crc/CRC.html

CRC for Intelligent Manufacturing Systems and Technologies www.crcimst.com.au

CRC for Cast Metals Manufacturing www.cast.crc.org.au

CRC for International Food Manufacture and Packing Science www.foodpack.crc.org.au

CRC for MicroTechnology www.microtechnologycrc.com

CRC for Construction Innovation Ph. 07 3864 4108

CRC for Functional Communication Surfaces Ph. 03 9905 3456

CRC for Innovative Wood Manufacturing Ph. 03 5321 4113

CRC for Railway Engineering and Technologies Ph. 07 4930 9549

> AGRICULTURE AND RURAL BASE MANUFACTURING

CRC for Sustainable Production Forestry www.forestry.crc.org.au

CRC for Tropical Plant Protection www.tpp.uq.edu.au

CRC for Viticulture www.winetitles.com.au/crcv

CRC for Cattle and Beef Quality www.beef.crc.org.au

Australian Cotton CRC www.cotton.pi.csiro.au

CRC for Sustainable Sugar Production www-sugar.jcu.edu.au

CRC for Molecular Plant Breeding www.molecularplantbreeding.com

CRC for Sustainable Rice Production www.ricecrc.org

Australian Sheep Industry CRC Ph. 02 6776 1301

CRC for Innovative Dairy Products Ph. 03 9602 5300

CRC for Sustainable Aquaculture of Finfish Ph. 03 6211 6666

CRC for Value Added Wheat hwarwick@wheatcrc.csiro.au

AUSTRALIAN SCIENCE

> MINING AND ENERGY

CRC for Mining Technology and Equipment www.cmte.org.au
GK Williams CRC for Extractive Metallurgy www.proceng1.chemeng.unimelb.edu.au/gkw.html
Australian Petroleum CRC www.apcrc.com.au
AJ Parker CRC for Hydrometallurgy www.parkercentre.crc.org.au
CRC for Clean Power from Lignite www.cleanpower.com.au
Australian CRC for Renewable Energy www.acre.murdoch.edu.au
CRC for Coal in Sustainable Development www.newcastle.edu.au/department/black_coal_crc/
CRC for Landscape Evolution and Mineral Exploration www.leme.anu.edu.au
CRC for Predictive Mineral Discovery Ph. 08 9389 8421

> MEDICAL SCIENCE AND TECHNOLOGY

CRC for Tissue Growth and Repair www.crc-tgr.edu.au
CRC for Cellular Growth Factors www.ludwig.edu.au/crc-cgf
CRC for Eye Research and Technology www.unsw.edu.au/crcert/CRCERT01.HTM
CRC for Cochlear Implant and Hearing Aid Innovation www.medoto.unimelb.edu.au/crc
CRC for Vaccine Technology www.crc-vt.qimr.edu.au
CRC for Aboriginal and Tropical Health www.ath.crc.org.au
CRC for Discovery of Genes for Common Human Diseases www.genecrc.org
CRC for Asthma www.asthma.crc.org.au
CRC for Chronic Inflammatory Diseases Ph. 03 8344 5480
CRC for Diagnostics Ph. 07 3864 4015

COOPERATIVE RESEARCH CENTRES

> INFORMATION AND COMMUNICATION TECHNOLOGY

CRC for Enterprise Distributed Systems Technology www.dstc.edu.au

Australian Photonics CRC www.photonics.com.au

CRC for Sensor Signal and Information Processing www.cssip.edu.au

Australian Telecommunications CRC www.atcrc.com

CRC for Satellite Systems www.crcss.csiro.au

CRC for Smart Internet Technology www.crcsit.com

CRC for Technology Enabled Capital Markets Ph. 02 9299 1883

> ENVIRONMENT

CRC for Waste Management and Pollution Control www.crcwmpc.com.au

CRC for Antarctica and the Southern Ocean www.antcrc.utas.edu.au

CRC for Catchment Hydrology www.catchment.crc.org.au

CRC for the Biological Control of Pest Animals www.pestanimal.crc.org.au

CRC Reef Research Centre www.reef.crc.org.au

CRC for Freshwater Ecology enterprise.canberra.edu.au/WWW/www-crcfe.nsf

CRC for Tropical Rainforest Ecology and Management www.rainforest-crc.jcu.edu.au

CRC for Conservation and Management of Marsupials www.newcastle.edu.au/marsupialcrc

CRC for Sustainable Tourism www.crctourism.com.au

CRC for Coastal Zone, Estuary and Waterway Management www.coastal.crc.org.au

CRC for Greenhouse Accounting www.greenhouse.crc.org.au

CRC for Australian Weed Management crcweeds@waite.adelaide.edu.au

CRC for Plant-based Management of Dryland Salinity Ph. 08 9380 2505

CRC for Tropical Savannas Management savanna.ntu.edu.au

CRC for Water Quality and Treatment www.med.monash.edu.au/epidemiology/crc

PART THREE
TOMORROW AND BEYOND

AUSTRALIAN SCIENCE

ETHICAL ISSUES

SCIENTIFIC ADVANCES ALWAYS BRING ETHICAL ISSUES WITH THEM; THE CHALLENGE IS TO MANAGE THE TWO TOGETHER, AS LEIGH DAYTON REPORTS

Scientific advances promise to revolutionise the way we live. But the new science brings new problems, and those new problems need new answers, which will only come with public debate. *Australian Science* spoke to two leading thinkers.

Associate Professor Helga Kuhse is Senior Honorary Research Fellow at the Centre for Human Bioethics at Monash University, in Victoria. She is an internationally respected bioethicist, and the author of numerous books, including *Should the Baby Live?*, which she co-authored with Professor Peter Singer.

The Honourable Justice Michael Kirby is a Justice of the High Court of Australia. He has a long-held interest in the nexus between legal and ethical issues. He has served on many international bodies, among them the Organization for Economic Cooperation (OECD) Committee on Privacy and Data Security and the International Ethics Committee of the Human Genome Organisation (HUGO).

ETHICAL ISSUES

Both were asked what they considered the key ethical issues facing scientists today.

>> KUHSE

One of the key ethical questions lies at the interface of genetics and new reproductive technologies. Not only is it possible to help the infertile become parents, but unprecedented advances in genetics also provide us with an ever-increasing ability to determine what those who come after us will be like.

The prevention of genetic disability has long been regarded as a desirable goal, and has been the primary motivation behind prenatal testing, and the reason for many abortions.

Today's prospective parents can seek to prevent the birth of a genetically affected child in various ways: pre-conception testing, the use of donor gametes or embryos, in vitro fertilisation (IVF), and abortion. They may also soon have recourse to gene therapy.

Given this, I believe that one of the big questions that must be faced, not only by scientists and parents, but by society as a whole, is: what should those who come after us be like, and how much reproductive freedom should parents have?

>> KIRBY

There are many ethical issues facing the public and scientists. Among all of those, the intellectual property question is of real significance.

The UNESCO Universal Declarations on Human Genomes and Human Rights set out many principles. One is the requirement for protection of intellectual property. Another is that the human genome, in its natural state, is part of the common heritage of humanity.

But how do you translate that into intellectual property law and practice? The pharmaceutical corporations and experimental scientists must spend considerable money and lots of time to get a drug on the counter. They won't do it unless they have some assurance of financial reward.

In contrast, developing countries are concerned that results from the HUGO will go into wrinkle research, rather than malaria immunity.

They are also concerned that profits derived from experiments on samples collected from people in developing countries will be used to the benefit of the developed world, and will be out of the financial reach of developing countries.

UNESCO is about to address the issue of how one gets a proper balance between the legitimate claims of business to intellectual property protection and the legitimate concern of peoples in poorer countries for their common heritage.

It is difficult to get agreement on the ethical foundations of policy and law in areas such as intellectual property and embryonic stem cell research. An important message is that not to make a decision is to make a decision: not to have a law is to give a green light.

Research and technology are making great strides forward, but we need to stay equally focused on the ethical issues raised by those great strides.

AUSTRALIAN SCIENCE

THE NEXT BIG QUESTIONS

> FUTURE SCIENCE
JULIAN CRIBB SCIENCE COMMUNICATOR & DIRECTOR, NATIONAL AWARENESS, CSIRO

Quite simply, the big questions of science demand a pooling of intellect. We will never truly understand how the human mind functions without an ingenious interleaving of biochemistry, psychology and sociology. And we will never explain how genes create the myriad variety of life until geneticists, biologists, mathematicians and behavioural scientists seriously put their heads together. Only when we have learnt to integrate our science will we be equipped to tackle some of Australia's—indeed the planet's—intractable problems. How do we extend the useful human life expectancy by overcoming degenerative and environmental disease? How will we feed and sustain a planetary population of nine billion by 2050, and so avoid global and regional chaos? And, here in Australia, how can we reincarnate 70-plus plant and animal species extinguished since European settlement and re-create sustainable habitats for as many of them as possible? The challenge ahead is a collective one.

> BIOTECHNOLOGY
PROFESSOR ALAN TROUNSON DEPUTY DIRECTOR, MONASH INSTITUTE OF REPRODUCTION AND DEVELOPMENT

Science is poised for another large leap forward that will incorporate both genomics and cell therapy in our strategies for prevention and treatment of disease and injury. The 'new medicine' will almost certainly involve embryo-derived stem (ES) cells. Since ES cells are immortal, we may be able to direct them into tissue types needed for cell therapy and scale them to help overcome degenerative, genetic and functional disorders or diseases. Can ES cells be used as universal donors for all patients with or without some modification? This also seems likely. Can ES cells or their derivatives be used for bioengineering organs, nerve bundles, skin or even new body parts? This may be more distant. Can these ES cells be used as vehicles for gene therapy? This is also a distinct possibility. Our discoveries may herald a new medicine or provide radical strategies for inducing self-cure by activating dormant multipotential stem cell lineages that appear to exist in many adult tissues.

THE NEXT BIG QUESTIONS

> PLANT SCIENCE AND AGRIBUSINESS
DR JIM PEACOCK CHIEF, CSIRO PLANT INDUSTRY

Plant science now knows the complete genome sequence of the broad leaf plant *Arabidopsis* and will soon have the complete sequence of rice, one the world's major food crops. These are major steps forward. But knowing the genome sequence is merely like having the words of a dictionary without understanding their meanings. The challenge ahead is to discover how each gene product interacts with each protein in every cell type in a plant. A few years ago such a goal would have sounded like science fiction: today it's a demanding but achievable objective. With this growing knowledge base we expect to help achieve the production necessary to feed the world's growing population. Ultimately the goal is to provide food security for all populations—food that is optimised for human health and that is produced without detriment to the environment, especially to production lands and their water supplies.

> IMMUNOLOGY
PROFESSOR EMERITUS SIR GUSTAV NOSSAL THE UNIVERSITY OF MELBOURNE

The next big challenges in immunology all concern the issue of how to turn science's stunning insights into the molecular and cellular workings of the immune system into practical results for human health. Vaccines are history's most cost-effective public health tools. But, as yet, we have no vaccine for HIV-AIDS or malaria and the BCG (*bacillus* Calmette-Guérin) we use against tuberculosis is insufficiently strong. While great hope attaches to the new concept of DNA vaccines—whereby the cells of the human body act as factories for antigens or vaccine molecules—how can we make such vaccines work better? Immunology is challenged similarly by the wide gamut of autoimmune diseases, where immune cells turn traitorous and attack the body, as in insulin-dependent diabetes or multiple sclerosis. How can we manipulate our immune systems to thwart such vicious civil warfare? Equally, how do we induce the powerful immune system to fight cancer? What molecules in or on the cancer cell could become targets for the immune cells? Alternatively, will the sharp tools known as monoclonal antibodies kill the small nests of cancer cells frequently remaining after surgery or chemotherapy? These are the big questions now facing the world's immunologists.

> HUMAN NUTRITION
PROFESSOR KERIN O'DEA DIRECTOR, MENZIES SCHOOL OF HEALTH RESEARCH

During the next 10 years nutrition researchers must address a number of urgent questions. Genetic modification of foods will have enormous implications for health, world food supply and the environment. Can we develop a sufficiently strong public sector capacity to balance private sector commercial interests? In a similar vein, public health nutrition is a key component of public health. Can we develop a robust regulatory framework that both provides clear boundaries for industry—by nutrition monitoring and surveillance—and maintains long-term public health? We also need to understand the increasingly complex relationships between diet and disease. New bioactive phytochemicals are being discovered constantly. What are their implications for diet and disease and how should nutritionists communicate this complex interaction in a balanced manner? Nutrition research must also play a part in assessing the true cost of diet-related chronic diseases and how diet may assist the prevention and treatment of chronic human disease.

AUSTRALIAN SCIENCE

> WORLD HEALTH
PROFESSOR ROGER V. SHORT DEPARTMENT OF OBSTETRICS AND GYNAECOLOGY, UNIVERSITY OF MELBOURNE

How are we going to cope in a world whose human population will inevitably grow from the six billion of today to 9–10 billion by 2050? Almost all of the increase will occur in developing countries, which are destined to become poorer and poorer as a result. Sadly, the inexorable spread of HIV infection may act as a brake on population growth; it is already reducing life expectancy in many countries in sub-Saharan Africa. During the next 50 years, the epicentre of the pandemic will inevitably shift to India, which could be decimated by the disease. Meanwhile, the world's increasingly affluent developed countries will go into negative growth as the economic costs of higher education and career opportunities for women act as powerful disincentives to childbearing. We have the infrastructure to keep AIDS at bay, but can we be persuaded to use our ever-increasing wealth to help the disadvantaged billions who live in the developing world? That is the question.

> EPIDEMIOLOGY
PROFESSOR FIONA STANLEY DIRECTOR, TVW TELETHON INSTITUTE FOR CHILD HEALTH RESEARCH

One of the world's most vexing questions is how to discover solutions for our emerging public health problems. Sadly, modern epidemiology is failing public health. As a science, epidemiology arose in response to 20th century epidemics of chronic diseases, such as cardiovascular disease and lung cancer, most of which result from exceedingly complex causal factors. Unfortunately, epidemiologists have persisted with the notion that pinpointing single independent risk factors in individuals can identify the roots of such diseases. They seem to have forgotten that understanding cause—and thus prevention—in a population demands careful assessment of the historical and social context in which risk factors arise. Preventive strategies based on individual risk factors in epidemiology have failed and will continue to fail. As we face new epidemics of complex diseases, mental health and psychosocial problems, it is vital that we amend our causal paradigms and preventive strategies urgently. The big question is—how to develop a new epidemiology?

> HUMAN EVOLUTION
DR ALAN THORNE DEPARTMENT OF ARCHAEOLOGY AND NATURAL HISTORY, AUSTRALIAN NATIONAL UNIVERSITY

Humans stand on the brink of a truly new understanding of how we evolved. As the vast Human Genome Project now promises stunning new insights into our genetic make-up—and our evolution in the future—so the increasingly sophisticated understanding of genetics is unearthing secrets of our ancient past. The question of modern human origins is becoming much clearer as we uncover more and more ancient DNA from human remains in an increasing range of environments. Likewise, the emergence of really early human genetic material from as far back as six million years ago is facilitating far more precise timing of critical evolutionary events, such as the split between chimpanzees and humans. Now we wait for new dating techniques that will allow us to place human remains and behaviour in an even clearer sequence. Only then will we begin to truly resolve our understanding of species, comparative anatomy and human evolution.

> PHOTOVOLTAICS
PROFESSOR MARTIN A. GREEN PHOTOVOLTAICS SPECIAL RESEARCH CENTRE, UNIVERSITY OF NEW SOUTH WALES

Two big questions challenge photovoltaic science: how to get the cost down as quickly as possible and what will be the winning solar cell technology in the coming decade. Cost reduction is a problem of volume. Government-subsidised use of solar cells on private homes in urban areas is encouraging investment in new solar cell production facilities and technologies that are driving costs down. Even with subsidies, solar cell costs cannot currently compete with electricity from large coal plants, although home owners still purchase this type of system, with about 3 million expected to be installed worldwide by 2010. As to the eventual winning solar cell technology, I think the silicon-based thin-film technology being developed in Sydney will be a clear winner in the medium term. In the longer term, cell conversion efficiency will be a key factor. We are now pursuing new approaches to converting sunlight to electricity at close to the theoretical efficiency limit of 95%, compared to about 15% for the best present commercial product.

> ASTRONOMY
PROFESSOR RONALD D. EKERS DIRECTOR, CSIRO AUSTRALIA TELESCOPE FACILITY

The truly grand challenges ahead concern the entire universe. With the next generation of sensitive optical, infrared and radio telescopes, astronomers will attempt to see the formation of the first structures in the universe—the first galaxies and the first stars—and their subsequent evolution. We will try to identify mysterious 'dark matter', the dominant form of matter in the universe. Likewise, a range of black holes from stellar sizes to giants devouring galaxies almost certainly exist, and as we pin down their properties they will let us access extreme regimes of density and new areas of physics. Even more mysterious and energetic than black holes are gamma-ray bursters, multiwavelength fireworks that light up the universe roughly once a day. With the advent of huge optical and radio telescopes and far more powerful space observatories, astronomers will explore these celestial phenomena and redouble efforts to answer the two big questions: how did the universe form, and how did stars and planetary systems capable of sustaining life evolve?

> CAPITALISING ON OUR INTELLECT
DR ROBIN BATTERHAM CHIEF SCIENTIST, COMMONWEALTH OF AUSTRALIA

The urgent question now facing Australian science is how to turn intellectual capital into national assets. Universities are vital elements of a productive, knowledge-based economy. By teaching, universities spawn intellectual prowess. By researching, they stimulate start-up and spin-off companies that add to Australia's wealth. When science can provide the data, and technologists and engineers the skills to create solutions to national problems, the esteem in which scientists are now held will soar. The question is, can we establish a framework in which scientists move with ease and excitement from 'blue sky' research to 'down-to-earth' practicality? For the moment, science for public good will continue to be funded. But until its results are linked more with community benefit I doubt that funding levels will increase. However, scientists who do take up the challenge of turning intellectual capital into national assets will contribute indelibly not just to science, but to the wellbeing of the entire country.

PART FOUR
VISION STATEMENTS

AUSTRALIAN SCIENCE

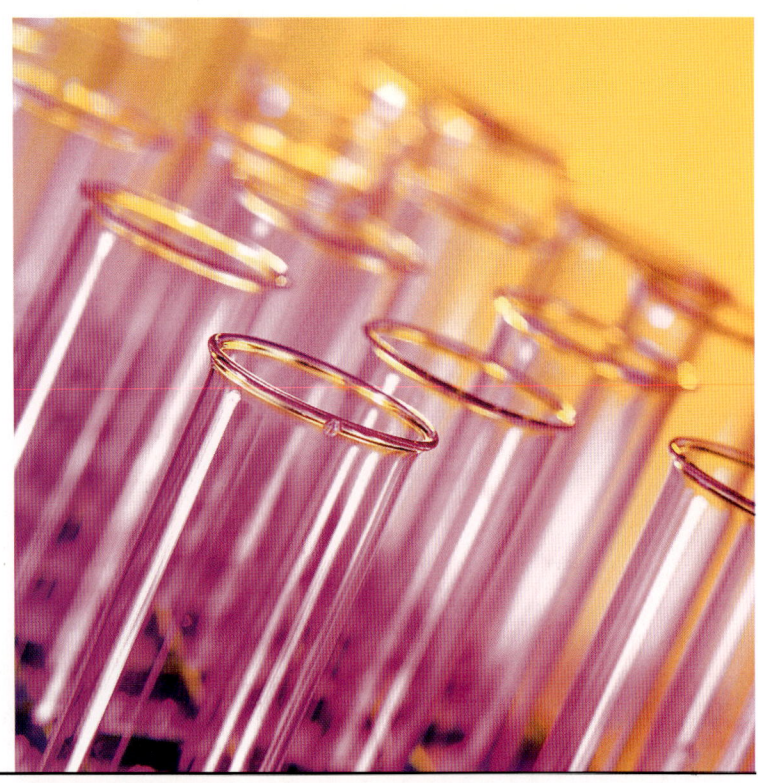

EFFECTIVE TREATMENTS

ASTRAZENECA IS A LEADING PHARMACEUTICALS MANUFACTURER AND RESEARCHER, WITH A GREAT COMMITMENT TO COLLABORATIVE WORK AND SOUND CUSTOMER RELATIONSHIPS

AstraZeneca is committed to improving quality of life through the development of innovative and effective healthcare solutions. Its leading position in many important areas of medicine enables AstraZeneca to make a difference to the lives of patients and the healthcare professionals who treat them.

Managing Director Jeays Lilley said the company strives to be first with new and innovative ideas in all areas of its business.

'We are privileged to have the opportunity to make a difference to human health, and we take our obligation seriously,' Lilley said.

'Our strategy is clear. We want to grow the value of the company, through the application of science and technology, by giving healthcare providers and their patients a continuous flow of new and effective treatments for disease.'

AstraZeneca is Australia's biggest supplier of prescription pharmaceuticals and one of the world's leading pharmaceutical companies.

Much of their research and development (R&D) effort directly benefits the future health and wellbeing of Australians. In the past three years they've spent $42 million in Australia on research into tomorrow's medical cures, and their track record shows their expertise at finding solutions and coming up with breakthroughs.

VISION STATEMENT

AstraZeneca has a strong research base and a comprehensive product portfolio designed to fight disease in seven areas of medical need. Their products in gastrointestinal medicine and anaesthesia lead the market. They are one of the foremost companies in oncology (cancer), respiratory and cardiovascular medicine. Their R&D activities also focus on infection control and diseases of the central nervous system.

This focus enables the company to build up teams with scientific, medical, regulatory and customer expertise in specific disease areas, thereby capitalising on the breadth of experience of its workforce and its existing product base.

AstraZeneca employs around 900 people in Australia and has a long history of close collaboration with local research institutions, universities and hospitals as well as with healthcare professionals and government authorities. The Australian operation plays an important role in the company's international network of R&D centres which involve more than 10,000 researchers.

'Our infrastructure here in Australia is sufficient to invite significant investment from our UK parent,' Lilley said. 'One of our largest collaborations is the joint R&D venture with the Griffith University in Brisbane. This is a key project in which thousands of plants and marine organisms are being screened for biologically active compounds that may be the medicines of the future.

'We are also one of the partner companies in the CRC for Asthma and we're currently in the process

> **ABOVE:** AstraZeneca boasts world class sterile production facilities at its North Ryde, NSW manufacturing base.

of applying for involvement in another major Australian CRC.'

AstraZeneca's R&D is focused generally on chronic disease states such as hypertension and cardiovascular, gastrointestinal and respiratory illnesses. An emerging area of focus is central nervous system diseases such as schizophrenia and depression, and the company is also developing drugs for the treatment of cancers.

'Strategically, this means we have a mix of products in a range of therapeutic areas, so we are not reliant on one stream or another,' Lilley said.

While their high-profile basic research program continues at Griffith University, a network of teaching hospitals, physicians and paramedical staff continue clinical research studies of AstraZeneca pharmaceuticals in every Australian capital city and in regional centres including Port Lincoln, Ballarat, Dubbo and Frankston.

AUSTRALIAN SCIENCE

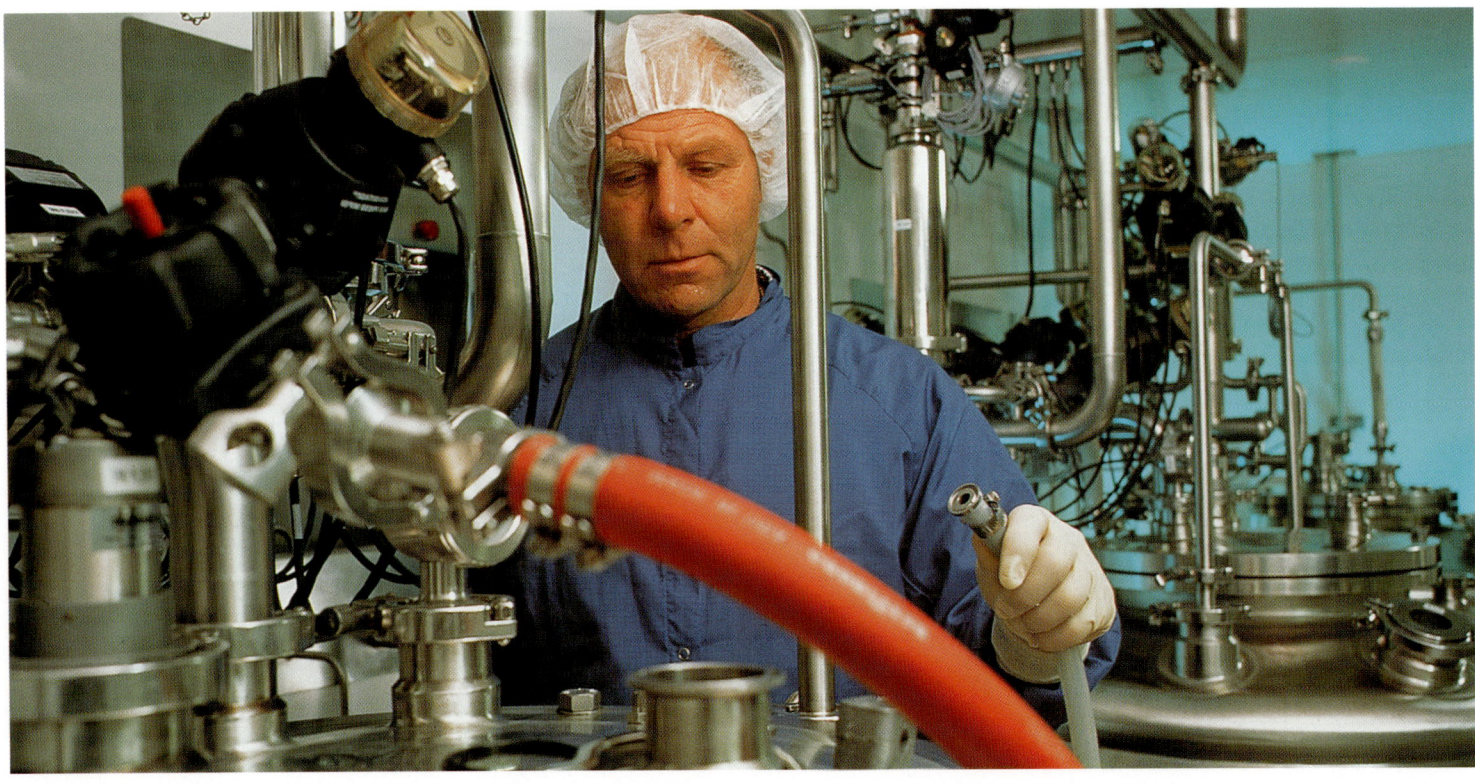

> **ABOVE:** AstraZeneca aspires to be perceived as the pharmaceutical employer of choice, from the factory floor to the boardroom.

AstraZeneca is committed to finding medical cures for the future, and the company takes that commitment very seriously. It's clear that such an emphasis on R&D can't take place without significant investment.

AstraZeneca invested more money in research and development during 1999/2000 than any other pharmaceutical company worldwide, according to the latest Department of Trade and Industry study.

Despite undergoing a merger, AstraZeneca's R&D expenditure rose to 15.8 per cent of sales, from 12.8 per cent the previous year. The company invests substantially in Australia because it believes in the long-term viability of a strong local research-based pharmaceutical industry. The budget allocated for R&D activities in Australia during 2000 is nearly $19 million.

Other investments in capital and infrastructure in Australia amount to more than $150 million. On top of this, the company recently announced a further $30 million investment in their leading-edge manufacturing plant in Sydney.

AstraZeneca Australia is one of the key sites in the global AstraZeneca supply chain. In the past four years they have invested more than $83 million in local infrastructure. In 1998 they opened a $68 million sterile pharmaceutical manufacturing facility in Sydney which is acknowledged as one of the most high-tech in the world. Australia and 17 other countries are supplied with medicines made in this state-of-the-art plant, with exports rising from less than $10 million in 1992 to more than $90 million in 1999.

Leading AstraZeneca products currently on the Australian market include treatments for reflux and ulcers (Losec®), high blood pressure (Zestril®, Plendil®, Tenormin®, Betaloc® and Atacand®), asthma (Pulmicort®, Oxis®), cancer (Arimidex™,

VISION STATEMENT

Zoladex®), as well as general anaesthetics (Diprivan®) and local anaesthetics (Xylocaine® and Naroin®).

In addition to their existing stable of world-leading products, AstraZeneca has a substantial portfolio of new chemical entities which are progressively coming to market.

'We are spending US$2.4 billion dollars on R&D and the development of collaborations with universities and researchers from around the world is giving us a large competitive advantage,' Jeays Lilley said.

'As a result, we have a whole range of new products becoming available to Australians.

position for the future. In order to achieve their goals—one of which is to become the first billion-dollar pharmaceutical company in Australia—there are a number of qualitative and collaborative changes taking place.

The growth of the business through the next decade will be fuelled by ensuring optimal market shares for AstraZeneca's range of existing products, coupled with successfully introducing of a range of new products.

Increased efforts are being devoted to the application of leading-edge science and technology to improve the quality and efficiency of their drug

ASTRAZENECA AIMS TO ATTRACT AND RETAIN THE BEST TALENT BY BUILDING A CULTURE WHICH VALUES, RECOGNISES AND REWARDS OUTSTANDING PERFORMANCES IN ALL ASPECTS OF ITS BUSINESS

The first of these products—a medication for schizophrenia called Seroquel®—was launched recently in Australia and is now available on the pharmaceutical benefits scheme.

'Several other new central nervous system products—including treatments for depression, bipolar disorders, anxiety disorders, movement disorders, dementia, acute stroke, epilepsy and multiple sclerosis—are in various stages of development.

'We are also set to launch another new product next year for treating acid-related gastrointestinal disorders. Looking to the future, our pipeline has been somewhat enhanced by the joining of the two companies in our recent merger, so we have a lot more planned to follow.'

Jeays Lilley said AstraZeneca has a visionary

discovery process, thereby providing a strong flow of high-potential candidates for development as new medicines.

The company aims to attract and retain the best talent by building a culture which values, recognises and rewards outstanding performance in all aspects of its business. Future plans for AstraZeneca Australia also include greater involvement in early-stage clinical research, known as Phase 2 and Phase 3 trials.

'We express a high level of social conscience in contributing to Australia,' Jeays Lilley said.

'We're doing our bit for employment here and we're exploring areas of common good. We see a responsibility to continue our present leadership in several parts of the pharmaceutical industry in Australia.'

AUSTRALIAN SCIENCE

RESEARCH LEADERSHIP

THE UNIVERSITY OF SYDNEY, NOW 150 YEARS OLD, IS ENTERING THE 21ST CENTURY AS AN INTERNATIONALLY RECOGNISED LEADER IN BOTH TEACHING AND RESEARCH

The University of Sydney, Australia's first university, was established in 1850. Throughout its history, it has been recognised as a provider of outstanding teaching and a centre of research excellence.

As it enters the new millennium, the University has a commitment to strengthen its position as a major research-intensive university, an outstanding provider of high-quality teaching, and a university of high standing in the international community of scholars.

When teaching began in 1852, there were just three professors and 30 students. The University now has more than 2,200 full-time academic staff and 36,000 students, giving it the best ratio of staff to students of any major Australian university. More than 215,000 students have graduated since the university's foundation.

The University's 18 faculties and are organised into three colleges—Health Sciences, Humanities and Social Sciences, and Sciences and Technology. Together, these colleges offer a wide range of courses in teaching departments spread across 10 campuses.

This diversity of teaching and depth of scholarship is matched only by the breadth and excellence of the University's internationally recognised research.

VISION STATEMENT

>> INTENSIVE RESEARCH EFFORT

Nourished by rich traditions, the University of Sydney is committed to funding research initiatives that maintain its leadership position and prepare the way for continued research excellence into the 21st century.

The University currently houses a prestigious Australian Research Council (ARC) special research centre—examining the ecological impacts of coastal cities, and it is home to four ARC key centres for teaching and research. These are in: microscopy and microanalysis, transport management, field robotics and polymer colloids. It has also won the lion's share of this year's ARC research grants, with $13.62 million awarded for new research in 2001.

The University outperformed every other university in the country in three out of the four major ARC schemes: large grants, fellowships and the research infrastructure, equipment and facilities program.

To further develop its reputation as an institution where pure and applied research and training are conducted to nationally and internationally recognised standards, the University has also designed a range of internal funding mechanisms to meet the needs of researchers in different disciplines. These include funding allocated according to research performance and for the direct support of research and research students.

>> INDUSTRY LINKS AND INTERNATIONAL TIES

The University's major strategy for ensuring its future success are to identify and support nodes of internationally recognised research

> **ABOVE:** Cutting-edge research at Sydney—Australia's leading research university.

>> ROLES AND VALUES

THE ROLE OF THE UNIVERSITY IS TO CREATE, PRESERVE, TRANSMIT AND APPLY KNOWLEDGE THROUGH TEACHING, RESEARCH, CREATIVE WORKS AND OTHER FORMS OF SCHOLARSHIP

In carrying out this role, the University reaffirms its commitment to:

> Institutional autonomy, recognition of the importance of ideas, intellectual freedom to pursue critical and open inquiry, and social responsibility
> Tolerance, honesty and respect as the hallmarks of relationships throughout the University community
> Understanding the needs and expectations of those whom it serves and striving constantly to improve the quality and delivery of its services.

AUSTRALIAN SCIENCE

> **ABOVE:** Sydney University students—part of the international community of scholars.

and to identify and foster areas of strategic priority.

As part of an international community of scholarship, the University of Sydney realises that its institutional goals cannot be achieved by acting in isolation. Existing overseas links will be intensified, involving students and academics from a full range of countries and cultures, to provide the best opportunities possible for the University's students and staff.

While flexibility and rationalisation of course offerings between the University of Sydney and other like-minded institutions is likely to continue as part of a healthy education system, there will still be keen international competition for the best students and staff, and for access to research funding.

The University will continue to develop long-term relationships with industry through the cooperative research centre program, through collaborative and contract research and through consultancies. These relationships are not only important in terms of the University's mission, but also serve to underscore the relationship between the institution and the society within which it is embedded.

>> KEY STRENGTHS FOR THE FUTURE

The University has a number of formally identified areas of institutional strengths, including cellular and molecular biology and biotechnology, marine and ocean research, advanced materials, environmental and ecological sustainability, astrophysics, cancer, cardiovascular and respiratory disease, neuroscience, information and communication

VISION STATEMENT

technology, pure mathematics, ageing health and disability, microbiology, infectious diseases and immunology, and public health and health service research.

These institutional strengths are broad areas that often cross disciplinary boundaries and are underpinned by a number of major research efforts. This identification of institutional strengths has been instrumental in targeting and developing extensive and innovative research programs in key focus areas.

There are numerous departments in the University, spanning a wide range of interests. Each is characterised by the pursuit of research interests and other scholarly activities aimed at advancing the fund of human knowledge.

The University of Sydney offers the largest research higher degree program in Australia. The rich diversity of courses offered, including a wide variety of cross-disciplinary degrees and combinations of degrees, attracts high-quality students from around the country and across the world.

The proportion of international students is increasing steadily, with the University now ranking sixth in terms of on-campus international students. To encourage this trend, the University has developed innovative teaching strategies and made major commitments to flexible learning and the use of technology in teaching.

In order to pave a solid path into the new millennium, the University of Sydney is highly focussed on sound management and quality learning.

That means students learn alongside staff who are themselves engaged in creative scholarship and learning. It means that general and academic staff work together in achieving the University's core objectives. It means everyone has space to make their individual contribution and to sense how they form part of a shared and purposeful enterprise.

>> MAJOR GOALS IN THE NEW MILLENNIUM

The University of Sydney will:
> Maintain and enhance its position as an outstanding provider of high-quality undergraduate and postgraduate teaching, both in Australia and internationally
> Continue to provide access to tertiary study and appropriate support for students from a diversity of backgrounds
> Develop its reputation as an institution where pure and applied research and research training are conducted to nationally and internationally recognised standards, including research relevant to the economic, social and cultural wellbeing of Australia and the region
> Further enhance its position as a university of high standing in the international community of scholars
> Continue to make a significant contribution to the wellbeing and enhancement of the wide range of professions with which it engages
> Improve its position as an efficient, effective and responsible institution, striving to meet the needs of students and staff, and committed to quality in all aspects of its operations
> Maintain and enhance its position as a leading contributor to the opinions and ideas, cultures and lifestyles of the many communities it serves locally, nationally and internationally, via the provision of knowledge, opportunity and encouragement.

AUSTRALIAN SCIENCE

INVESTING IN R&D

MERCK SHARP & DOHME IS COMMITTED TO BEING AN INNOVATIVE AND VIGOROUS PLAYER IN THE KNOWLEDGE-BASED ECONOMY AND AN INDUSTRY LEADER IN IMPROVING HEALTH OUTCOMES.

Merck Sharp & Dohme (Australia) Pty Limited (MSD) has enjoyed a long and rich association with Australian medical research dating back to Howard Florey, when Merck joined with other companies in manufacturing penicillin for the Allied forces in World War II. Florey's legacy lives on, with millions of lives saved due to Florey and his co-collaborators' work in discovering penicillin, and by pharmaceutical manufacturerers' ability to transform a new chemical entity (NCE) into a safe and effective medicine.

The pharmaceutical industry is a crucial part of the scientific community and the largest funder of medical research in Australia, investing more than A$300 million annually in the search for new and improved therapies.

Merck Sharp & Dohme is among the nation's top 40 firms investing in R&D and spending millions of dollars in original research and development and clinical trials of new medicines in Australia. Other key projects MSD supports include:

> The development of a vaccine for the human papilloma virus, associated with more than 93 per cent of cervical cancer cases;
> A six-year study of hypertension in 6,000 elderly patients, aimed at identifying optimal hypertension therapy;
> The Cooperative Research Centre for Asthma and the Australian Diabetes Survey, which the federal government is also partnering.

In recent years MSD has also supported:

> Research at the Garvan Institute of Medical Research into the genetic causes of manic depression; and
> AMRAD Corporation—through a joint venture in AMRAD Pharmaceuticals since 1988. MSD recently acquired all of AMRAD Pharmaceuticals, and will continue to support the former parent company's research efforts for years to come through licensing the AMRAD name.

Research is the cornerstone of Merck & Co Inc globally. In 2000 Merck Research Laboratories spent US$2.4 billion (>A$3.2 billion) on R&D, underpinning the richest pipeline in the company's history. This pipeline promises to yield therapeutically unprecedented medicines, with NCEs in development for depression, HIV, cancer, heart disease, diabetes, infectious diseases, osteoarthritis, rheumatoid arthritis and pain.

Merck Sharp & Dohme is committed to being an innovative and vigorous player in the knowledge-based economy. We aim to lead the industry in improving health outcomes and realising the economic benefits of a flourishing pharmaceutical industry.

VISION STATEMENT

FOSTERING INNOVATION

CSL IS AN AUSTRALIAN HEALTHCARE PRODUCTS COMPANY THAT FOCUSES ON RESEARCH AND DEVELOPMENT, QUALITY ASSURANCE, A SKILLED AND INNOVATIVE WORKFORCE AND MARKETING ALLIANCES

CSL is an Australian company specialising in the development, manufacture and marketing of biologically based healthcare products. Not only is CSL the largest investor in Australian pharmaceutical research and development, it is also one of the largest employers in the local pharmaceutical manufacturing industry.

Its healthcare output includes life-saving products derived from human plasma, pharmaceuticals and diagnostics essential to community health, and animal health vaccines and diagnostics to protect livestock and companion animals.

Heading into the new millennium, the company will firstly maintain its emphasis on meeting customers' expectations by supplying quality products and excellent service. CSL's commitment to developing a flexible, committed and skilled workforce which is rewarded for excellence and innovation will help them reach this goal.

There will also be ongoing investment in the development of new products and the pursuit of collaborations that build on the company's existing scientific, manufacturing and marketing expertise.

CSL will also expand its business by actively marketing products nationally and internationally.

CSL's innovation and new product development are a major driving force behind its growth.

The overriding goal is to deliver novel ideas to the market. CSL achieves this by adding value in late-stage research, early-stage development, clinical trials, product registration and market entry.

The strategy is to build an intellectual property portfolio that is derived from both internal innovation and partnering with the academic community.

As part of this program, CSL is currently working on a number of products that act on, or through, the immune system. The company's portfolio consists of vaccines and immunotherapies aimed at two primary targets—infectious disease and cancer. CSL is the industry partner in the government-funded Cooperative Research Centre for Vaccine Technology.

The company also has many close associations with universities and medical research institutes currently working on early-stage projects.

The company plans to build a global business for its bioplasma products, with strong market positions in both Europe and the United States. It is also striving to improve its market position in Australia and build on its competitive strengths.

Most importantly, in every area, innovation will be fostered via new product development and partnerships with collaborators to ensure ongoing growth and competitiveness.

AUSTRALIAN SCIENCE

Collins Class Submarine. PICTURE: NAVY PHOTOGRAPHIC UNIT

CHARTING NEW WATERS

THOMSON MARCONI SONAR IS ONE OF THE WORLD'S LARGEST UNDERWATER DEFENCE EQUIPMENT DESIGNERS AND MANUFACTURERS, AND A LEADING SUPPLIER OF ACOUSTIC SENSORS AND SONAR SYSTEMS

Thomson Marconi Sonar has more than 50 years experience in undersea warfare and is clearly positioned as the national leader in underwater warfare in Australia, France and the United Kingdom.

Thomson Marconi Sonar in Australia maximises the strengths of local research while maintaining access to an international research and development capability. This combination of local expertise and international access ensures the best result for Australia.

Their international reputation is highlighted by the fact that nearly 50 navies from around the world currently utilise their equipment and expertise. In Australia alone they have supplied advanced and highly capable sonar systems to the Royal Australian Navy's COLLINS submarines, ANZAC and FFG frigates, and the new HUON class Minehunting Ships, to name a just few.

Thomson Marconi Sonar's mission is simply stated: to be the market leader through the provision of sonar products and services which customers trust to meet or exceed their expectations.

The company is achieving this mission by creating a substantial operations base in Australia to nurture and develop local technologies, with access to international technology and expertise as required. This mix ensures that the current needs

VISION STATEMENT

of the Australian defence forces are well met and paves the way for their future needs.

Thomson Marconi Sonar in Australia designs, manufactures and supports sonar systems for Australia and New Zealand, as well as for export to international markets.

The main area of business in Australia is in defence, but civil operations such as the marine seismic industry are a growing part of their business, accounting for around 45% of sales turnover in 2000.

They also specialise in systems engineering for underwater acoustic systems, and maintain a high level of expertise in acoustics and signal processing.

Their capabilities in mechanical, electrical and thermal modelling are extensive in the field of sonar performance prediction.

Yet the products developed by Thomson Marconi Sonar are just the visible tip of their wide-ranging abilities. Because research and development is such a major part of the company's business strategy, a mountain of expertise lies behind their products and services. Expertise which has firmly placed Thomson Marconi Sonar at the leading edge of innovation.

Thomson Marconi Sonar recognises that customer trust and satisfaction are vital to their future success, so they provide products and services that are highly reliable, of high quality, and delivered on schedule.

When it comes to work practices, Thomson Marconi Sonar provides sound methodologies for every project to ensure that the end result is of the highest quality and more than meets their customers' needs. Their broad range of sonar products and vast local and international experience ensures that customers can have the high level of confidence required in the significant services they provide to the Australian community and the local economy.

Thomson Marconi Sonar's customer relationships continue throughout the life of a product. With an average product age of around 25 years, it's vital that these business relationships are successful. Their future success is ensured by the fact that they guarantee maximum product performance for their customers throughout a product's life.

Innovation is important to meet the evolving needs of technologically advanced industries such as defence and oil and gas exploration. To retain their ability to satisfy the latest demands of a changing market, continued and extensive research and development is integral to their future plans.

The engineering team at Thomson Marconi Sonar includes the best technical experts available. Their fields of expertise and skills cover electronic and mechanical hardware, acoustics, ceramics and materials science, signal processing, software engineering, production engineering and project management.

Such skills and expertise are an essential component of the company's future development. Thomson Marconi Sonar is striving to develop new and innovative technologies that position them well for high performance and for the provision of world-leading technology in the future.

In the drive for innovation, their experts are constantly charting new waters and seeking the best technical solutions to create better products.

PART FIVE
DIRECTORY OF PARTICIPANTS

ABN. 49 132 008 380

See page 45 for our Case Study

ASSOCIATION OF AUSTRALIAN MEDICAL RESEARCH INSTITUTES

c/- Murdoch Childrens Research Institute, 10th Floor, Royal Children's Hospital,
Flemington Road, Parkville VIC 3052
Phone. **03 8341 6226**   Fax. **03 9348 1391**

morrellk@cryptic.rch.unimelb.edu.au

Prof Bob Williamson FRS **Director**
Anne Cronin **Chief Operating Officer**

AAMRI sponsorship has been a combination of these Medical Institutes leading the way in medical research in Australia and the world.

- Austin Research Institute
- Baker Medical Institute
- Centenary Institute of Cancer Medicine and Cell Biology
- Children's Cancer Institute Australia
- Garvan Institute
- Hanson Centre for Cancer Research
- Howard Florey Institute
- Macfarlane Burnet Centre for Medical Research
- Mater Medical Research Institute
- Menzies School of Health Research
- Murdoch Childrens Research Institute
- Prince Henry's Institute of Medical Research
- Prince of Wales Medical Research Institute
- St. Vincent's Institute of Medical Research
- The Mental Health Research Institute of Victoria
- The Walter and Eliza Hall Institute of Medical Research
- TVW Telethon Institute for Child Health Research

AUSTRALIAN SCIENCE

ABN. 62 291 911 396   See page 83 for our Case Study

ALLENS ARTHUR ROBINSON

Stock Exchange Building, Level 27, 530 Collins Street, Melbourne, VIC 3000
GPO Box 1776Q, Melbourne VIC 3001
Phone. **03 9614 1011**   Fax. **03 9614 4661**

arh@arh.com.au
www.arh.com.au

Michael Robinson/Kevin McCann **Chairman**
Tom Poulton **Managing Partner**

ABN. 61 426 486 715   See page 63 for our Case Study

ANTI-CANCER COUNCIL OF VICTORIA

1 Rathdowne Street, Carlton VIC 3053
Phone. **03 9635 5000**   Fax. **03 9635 5270**

enquiries@accv.org.au
www.accv.org.au

Robert C. Burton **Director**
David J. Hill **Director Cancer Control Research Institute**

ABN. 54 009 682 311   See pages 56 and 64 for our Case Study

ASTRAZENECA PTY LTD

Alma Road, North Ryde NSW 2113
PO Box 131, North Ryde NSW 1670
Phone. **02 9978 3500**   Fax. **02 9978 3700**

AstraZeneca.Information@AstraZeneca.com
www.AstraZeneca.com.au

Jeays Lilley **Managing Director**
Greg Williams **Chief Financial Officer**

DIRECTORY OF PARTICIPANTS

ABN. 26 007 418 224 See page 42 for our Case Study

AUSTIN RESEARCH INSTITUTE

Austin & Repatriation Medical Centre, Kronheimer Building, Studley Road, Heidelberg VIC 3084
Phone. **03 9287 0666**   Fax. **03 9287 0600**

i.mckenzie@ari.unimelb.edu.au
www.ari.unimelb.edu.au

Professor Ian McKenzie **Director**
Professor Mark Hogarth **Deputy Director**
Professor Mauro Sandrin **Deputy Director**

ARBN. 023 580 663 See page 62 for our Case Study

AUSTRALIAN PHARMACEUTICAL MANUFACTURERS ASSOCIATION

Level 7, 88 Walker Street, North Sydney NSW 2060
Phone. **02 9922 2699**   Fax. **02 9959 4860**

info@apma.com.au
www.apma.com.au

Jeays Lilley **Chairman**
Alan Evans **Chief Executive Officer**

ABN. 59 003 849 198 See page 76 for our Case Study

AUSTRALIAN PROTEOME ANALYSIS FACILITY

Level 4 Building F7B, Research Park Drive, Macquarie University, North Ryde NSW 2109
Phone. **02 9850 6201**   Fax. **02 9850 6200**

apafinfo@proteome.org.au
www.proteome.org.au

Professor Gary Cobon **Director**
Dr Brad Walsh **Facility Manager**

AUSTRALIAN SCIENCE

ABN. 51 452 193 160   See page 131 for our Case Study

AUSTRALIAN RESEARCH COUNCIL

AGSO Building, Corner Hindmarsh Drive and Jerrabomberra Avenue, Symonston ACT 2609
GPO Box 9880, Canberra ACT 2601
Phone. **02 6284 6600**   Fax. **02 6284 6601**

www.arc.gov.au

Professor Vicki Sara **Chair**

ABN. 15 092 808 850   See page 39 for our Case Study

AUSTRALIAN TECHNOLOGY PARK INNOVATIONS PTY LTD

Suite 145, National Innovation Centre, Australian Technology Park, Eveleigh, NSW 1430
Phone. **02 9209 4444**   Fax. **02 9319 3870**

www.atp.com.au

Professor Steve Bakoss **Chief Executive Officer**

ABN. 60 234 497 945   See page 51 for our Case Study

BAKER MEDICAL RESEARCH INSTITUTE

Commercial Road, Prahran VIC 3181
PO Box 6492, St Kilda Road Central, Melbourne VIC 8008
Phone. **03 9522 4333**   Fax. **03 9521 1362**

Baker@baker.edu.au
www.baker.edu.au

Norman O'Bryan **President of the Board BMRI**
John Funder AO **Director of BMRI**

DIRECTORY OF PARTICIPANTS

ABN. 28 006 479 081

See page 83 for our Case Study

BIOTA HOLDINGS LIMITED

Level 4, 616 St Kilda Road, Melbourne VIC 3004
Phone. **03 9529 2311**   Fax. **03 9529 2261**

info@biota.com.au
www.biota.com.au

Brian Healey **Chairman**
Hugh Niall **Chief Executive Officer**

ABN. 22 654 201 090

See page 50 for our Case Study

CENTENARY INSTITUTE OF CANCER MEDICINE AND CELL BIOLOGY

Building 93, Royal Prince Alfred Hospital, Missenden Road, Camperdown NSW 2050
Locked Bag 6, Newtown NSW 2042
Phone. **02 9565 6156**   Fax. **02 9565 6101**

sciman@centenary.usyd.edu.au
www.centenary.usyd.edu.au

Malcolm Noad **Chairman**
Antony Basten **Executive Director**

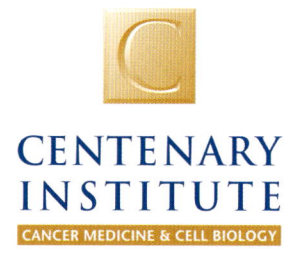

ABN. 41 072 279 559

See page 52 for our Case Study

CHILDREN'S CANCER INSTITUTE AUSTRALIA FOR MEDICAL RESEARCH

High Street, Randwick NSW 2031
PO Box 81, Randwick NSW 2031
Phone. **02 9382 1829**   Fax. **02 9382 1850**

info@ccia.org.au
www.ccia.org.au

Michelle Haber **Director**
Mark Franklin **General Manager**

AUSTRALIAN SCIENCE

ABN. 41 450 623 885   See page 109 for our Case Study

COOPERATIVE RESEARCH CENTRE FOR BLACK COAL UTILISATION
Advanced Technology Centre, University of Newcastle, Callaghan NSW 2308
Phone. **02 4921 7314**   Fax. **02 4921 7168**

black-coal@newcastle.edu.au
www.newcastle.edu.au/department/black_coal_crc

Ken Smith **Chairman**
John Hart **Executive Director**

ABN. 24 115 172 498   See page 115 for our Case Study

COOPERATIVE RESEARCH CENTRE FOR CLEAN POWER FROM LIGNITE
677 Springvale Road, Mulgrave VIC 3170
Phone. **03 9239 0800**   Fax. **03 9561 0710**

reception@cleanpower.com.au
www.cleanpower.com.au

David Brockway **Chief Executive Officer**
Pete Saunders **Chairman**

ABN. 99 051 588 348   See page 66 for our Case Study

CSL LIMITED
45 Poplar Road, Parkville VIC 3052
Phone. **03 9389 1911**   Fax. **03 9389 1434**

www.csl.com.au

Brian McNamee **Managing Director**
Colin Armit **President Pharmaceutical Group**

DIRECTORY OF PARTICIPANTS

ABN. 39 000 233 992   See page 43 for our Case Study

ELI LILLY AUSTRALIA PTY LIMITED
112 Wharf Road, West Ryde NSW 2114
Phone. **02 9325 4444**   Fax. **02 9325 4334**

www.lilly.com.au

Answers That Matter.

Nancy Lilly **Managing Director**
David Grainger **Director, Corporate Affairs and Health Economics**

ABN. 62 330 391 937   See page 44 for our Case Study

GARVAN INSTITUTE OF MEDICAL RESEARCH
384 Victoria Street, Darlinghurst NSW 2010
Phone. **02 9295 8100**   Fax: **02 9295 8101**

j.elliott@garvan.org.au
www.garvan.org.au

John Shine **Executive Director**

See page 55 for our Case Study

GLAXOSMITHKLINE AUSTRALIA
Pharmaceutical Division: PO Box 168, Boronia VIC 3155
Consumer Healthcare Division: Locked Bag 3, Ermington NSW 2115
Phone. **03 9721 6000** Pharmaceuticals/**02 9684 0888** Consumer Healthcare Division
Fax. **03 9729 5319** Pharmaceuticals/**02 9684 1018** Consumer Healthcare Division

www.gsk.com.au

Sue Middleton, **Director of Government & Public Affairs, Pharmaceutical Division**
Paul Brown, **Manager of Industry & Public Affairs, Consumer Healthcare Division**

AUSTRALIAN SCIENCE

ABN. 35 302 506 443   See page 57 for our Case Study

HANSON CENTRE FOR CANCER RESEARCH

Frome Road, Adelaide SA 5000
PO Box 14, Rundle Mall, Adelaide SA 5000
Phone. **08 8222 3033**   Fax. **08 8222 3035**

hanson@imvs.sa.gov.au
www.imvs.sa.gov.au/hanson

Professor John Gollan **Director**
Dr Christopher Juttner **Deputy Director**
Dr Howard Morris **Chief Operating Officer**

ABN. 13 773 626 855   See page 53 for our Case Study

HOWARD FLOREY INSTITUTE

Gate 11, Royal Parade, The University of Melbourne VIC 3010
Phone. **03 8344 5654**   Fax. **03 9348 1707**

h.deaizpurua@hfi.unimelb.edu.au
www.hfi.unimelb.edu.au

Frederick Mendelsohn **Director**
Graeme Chandler **General Manager**

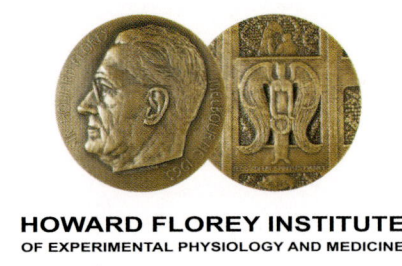

HOWARD FLOREY INSTITUTE
OF EXPERIMENTAL PHYSIOLOGY AND MEDICINE

ABN. 52 234 063 906   See page 65 for our Case Study

JOHN CURTIN SCHOOL OF MEDICAL RESEARCH

Mills Road, The Australian National University, Canberra ACT 2601
Phone. **02 6125 2589**   Fax. **02 6125 2337**

Peter.jeffrey@anu.edu.au
www.jcsmr.anu.edu.au

Judith Whitworth **Director**
Stephen Redman **Deputy Director**

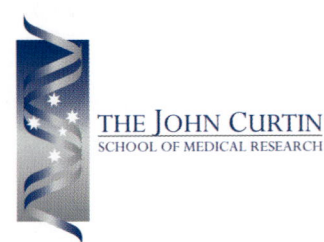

THE JOHN CURTIN
SCHOOL OF MEDICAL RESEARCH

DIRECTORY OF PARTICIPANTS

See page 46 for our Case Study

MACFARLANE BURNET CENTRE FOR MEDICAL RESEARCH

Yarra Bend Rd, Fairfield VIC 3078
PO Box 254, Fairfield VIC 3078
Phone. **03 9282 2111** Fax. **03 9282 2100**

www.burnet.edu.au

Professor John Mills **Director**
Mr Peter Spiller **Human Resources & Business Manager**

ABN. 94 710 251 744

See page 60 for our Case Study

MATER MEDICAL RESEARCH INSTITUTE

Level 3, Aubigny Place, Raymond Terrace, South Brisbane QLD 4101
Phone. **07 3840 2555** Fax. **07 3840 2550**

dhart@mmri.mater.org.au
www.mmri.mater.org.au

Prof Derek Hart **Director**
David Wood **Business Manager**

Mater Medical Research Institute

ABN. 70 413 542 847

See page 61 for our Case Study

MENZIES SCHOOL OF HEALTH RESEARCH

Building 58, Royal Darwin Hospital Compound, Rocklands Drive TIWI NT 0810
PO Box 41096, Casuarina NT 0811
Phone. **08 8922 8196** Fax. **08 8927 5187**

www.menzies.edu.au

Richard Ryan AO **Chair of the Governing Board**
Professor Kerin O'Dea **Director**

AUSTRALIAN SCIENCE

ABN. 14 000 173 508 See page 67 for our Case Study

MERCK SHARP & DOHME

54-68 Ferndell Street, South Granville NSW 2142
PO Box 79, Granville NSW 2142
Phone. **02 9795 9500** Fax. **02 9795 9595**

www.msda.com.au

Will Delaat **Managing Director**

ABN. 65 003 644 657 See page 37 for our Case Study

MICHAEL JOHNSON & ASSOCIATES PTY LIMITED

10 Darling Street, Balmain NSW 2041
Phone. **02 9810 7211** Fax. **02 9818 2297**

info@mjassoc.com.au
www.mjassoc.com.au

Kris Gale **Managing Director**
Andrew Hills **Senior Manager**
Chris Koutoulas **Senior Manager**

ABN. 12 377 614 012 See page 68 for our Case Study

MONASH UNIVERSITY

Australia Italy Malaysia South Africa United Kingdom
Phone. **+ 61 3 9902 6000** Fax. **+ 61 3 9905 4007**

inquiries@monash.edu.au
www.monash.edu.au

Jerry Ellis **Chancellor**
Professor David Robinson **Vice-Chancellor and President**

DIRECTORY OF PARTICIPANTS

ABN. 21 006 566 972   See page 82 for our Case Study

MURDOCH CHILDRENS RESEARCH INSTITUTE

Royal Children's Hospital, Flemington Road, Parkville VIC 3052
Phone. **03 8341 6200**   Fax. **03 9348 1391**

www.murdoch.rch.unimelb.edu.au

Prof Bob Williamson **Director**
Anne Cronin **General Manager**

ABN. 18 060 658 764   See page 70 for our Case Study

OPTISCAN PTY LTD

15-17 Normanby Road, Notting Hill VIC 3168
PO Box 1066, Mt Waverley MDC VIC 3149
Phone. **03 9538 3333**   Fax. **03 9562 7742**

info@optiscan.com
www.optiscan.com

Professor Ray Martin **Chairman**
Peter Delaney **Managing Director**

ABN. 52 780 433 757   See page 112 for our Case Study

PRICEWATERHOUSECOOPERS

201 Sussex Street, Sydney NSW 2000
GPO Box 2650, Sydney NSW 1171
Phone. **02 8266 0000**   Fax. **02 8266 9999**

www.pwcglobal.com/autice

Martyn Mitchell **Lead Partner**
—Technology, Information, Communication and Entertainment

AUSTRALIAN SCIENCE

ABN. 77 601 754 678   See page 49 for our Case Study

PRINCE HENRY'S INSTITUTE OF MEDICAL RESEARCH
Level 4, Block E, 246 Clayton Road, Clayton VIC 3168
PO Box 5152, Clayton VIC 3168
Phone. **03 9594 4372**   Fax. **03 9594 6125**

evan.simpson@med.monash.edu.au
www.med.monash.edu.au/phimr

Russell J. Fynmore AO **FCPA Chairman**
Professor Evan Simpson PhD **Director**

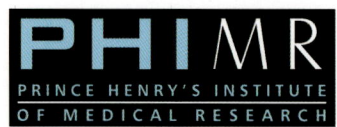

ABN. 94 050 110 346   See page 48 for our Case Study

PRINCE OF WALES MEDICAL RESEARCH INSTITUTE
Prince of Wales Hospital, Barker Street, Randwick NSW 2031
PO Box 82, St Pauls NSW 2031
Phone. **02 9382 2688**   Fax. **02 9382 2722**

powmri@unsw.edu.au
www.powmri.unsw.edu.au

William Penfold AM **Chairman**
Ian McCloskey AO **Executive Director**

ABN. 69 981 208 782   See page 79 for our Case Study

STATE GOVERNMENT OF VICTORIA
Science, Technology and Innovation Division, Department of State and Regional Development
GPO Box 4509RR, Melbourne VIC 3001
Phone. **03 9651 9283**   Fax. **03 9651 9236**

innovation@dsrd.vic.gov.au
www.innovation.vic.gov.au

Neil Edwards **Secretary**
Jane Niall **Executive Director, Science, Technology and Innovation**

DIRECTORY OF PARTICIPANTS

ABN. 52 004 705 640   See page 47 for our Case Study

ST. VINCENT'S INSTITUTE OF MEDICAL RESEARCH

9 Princes Street, Fitzroy VIC 3065
41 Victoria Parade, Fitzroy VIC 3065
Phone. **03 9288 2480**   Fax. **03 9416 2676**

j.martin@medicine.unimelb.edu.au
www.svimr.unimelb.edu.au

T. John Martin **Chief Executive Officer**
David Rees **Business Manager**

ABN. 20 078 532 180   See page 69 for our Case Study

STARPHARMA POOLED DEVELOPMENT LIMITED

343 Royal Parade, Parkville VIC 3052
Phone. **03 9662 7123**   Fax. **03 9662 7129**

info@starpharma.com
www.starpharma.com

Richard Oliver **Chairman**
Dr John Raff **Chief Executive Officer**

ABN. 76 437 529 296   See page 59 for our Case Study

THE MENTAL HEALTH RESEARCH INSTITUTE OF VICTORIA

155 Oak Street, Parkville VIC 3052
Locked Bag 11, Parkville VIC 3052
Phone. **03 9388 1633**   Fax. **03 9387 5061**

rjj@mhri.edu.au
www.mhri.edu.au

Dr Ben Lochtenberg **Chairman**
Professor David Copolov **Director**

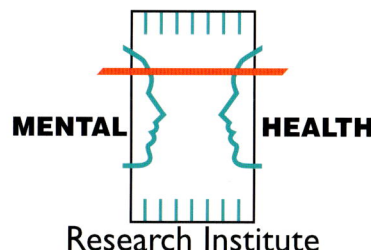

AUSTRALIAN SCIENCE

ABN. 31 411 813 344   See page 71 for our Case Study

THE QUEENSLAND INSTITUTE OF MEDICAL RESEARCH

The Bancroft Centre, 300 Herston Road, Brisbane QLD 4006
Post Office, Royal Brisbane Hospital QLD 4029
Phone. **07 3362 0222**   Fax. **07 3362 0111**

michaelS@qimr.edu.au
www.qimr.edu.au

Michael Good **Director**
Michael Staley **Chief Operating Officer**

ABN. 84 002 705 224   See page 38 for our Case Study

THE UNIVERSITY OF MELBOURNE

Grattan Street, Parkville VIC 3010
Postal. The University of Melbourne VIC 3010
Phone. **03 8344 6937**   Fax. **03 9347 6739**

g.swafford@unimelb.edu.au
www.unimelb.edu.au

Professor Alan Gilbert **Vice-Chancellor**
Professor Frank Larkins **Deputy Vice-Chancellor (Research)**

ABN. 15 211 513 464   See pages 72 and 97 for our Case Studies

THE UNIVERSITY OF SYDNEY

The University of Sydney, NSW 2006
Phone. **02 9351 2222**   Fax. **02 9351 5758**

vice-chancellor@vcc.usyd.edu.au
www.usyd.edu.au

Professor Gavin Brown **Vice-Chancellor**
Emeritus Professor Dame Leonie Kramer **Chancellor**

DIRECTORY OF PARTICIPANTS

ABN. 12 004 251 423   See page 58 for our Case Study

THE WALTER AND ELIZA HALL INSTITUTE OF MEDICAL RESEARCH

The Royal Melbourne Hospital, Royal Parade, Parkville VIC 3052
Post Office, The Royal Melbourne Hospital, VIC 3050
Phone. **03 9345 2555**   Fax. **03 9347 0852**

webmaster@wehi.edu.au
www.wehi.edu.au

Suzanne Cory **Director**
Margaret Brumby **General Manager**

ABN. 70 073 076 212

THOMSON MARCONI SONAR

274 Victoria Road, Rydalmere NSW 2116
Phone. **02 9848 3500**   Fax. **02 9848 3888**

marketing@tms-pty.com
www.tms-pty.com

Chris Jenkins **Managing Director**
Ashley Deacon **Marketing & Sales Director**

ABN. 86 009 278 755   See page 54 for our Case Study

TVW TELETHON INSTITUTE FOR CHILD HEALTH RESEARCH

100 Roberts Road, Subiaco WA 6008
PO Box 855, West Perth WA 6872
Phone. **08 9489 7777**   Fax. **08 9489 7700**

info@ichr.uwa.edu.au
www.ichr.uwa.edu.au

Fiona Stanley **Scientific Director**
Bruce McHarrie **Chief Financial Officer**

TVW TELETHON INSTITUTE FOR
Child Health Research

AUSTRALIAN SCIENCE

ABN. 84 059 048 770   See page 118 for our Case Study

COOPERATIVE RESEARCH CENTRE FOR **ADVANCED COMPOSITE STRUCTURES LTD**

506 Lorimer Street, Fishermans Bend, Victoria 3207

Phone. **03 9646 6544**   Fax. **03 9646 8352**

melbcrc@ozemail.com.au
www.crc-acs.com.au

Ken Harris **Chairman**
Ian Mair AM **Chief Executive Officer**

ABN. 60 847 010 598   See page 95 for our Case Study

COOPERATIVE RESEARCH CENTRE **FOR THE BIOLOGICAL CONTROL OF PEST ANIMALS**

Cnr Barton Highway and Bellenden Street, Crace ACT 2911

GPO Box 284, Canberra ACT 2601

Phone. **02 6242 1768**   Fax. **02 6242 1511**

office@pestanimal.crc.org.au
www.pestanimal.crc.org.au

Neil Willetts **Chairman**
Tony Peacock **Chief Executive Officer**

ABN. 41 450 623 885   See page 109 for our Case Study

COOPERATIVE RESEARCH CENTRE FOR **BLACK COAL UTILISATION**

Advanced Technology Centre, University of Newcastle, Callaghan NSW 2308

Phone. **02 4921 7314**   Fax. **02 4921 7168**

black-coal@newcastle.edu.au
www.newcastle.edu.au/department/black_coal_crc

Ken Smith **Chairman**
John Hart **Executive Director**

ABN. 24 115 172 498   See page 115 for our Case Study

COOPERATIVE RESEARCH CENTRE FOR **CLEAN POWER FROM LIGNITE**

677 Springvale Road, Mulgrave VIC 3170

Phone. **03 9239 0800**   Fax. **03 9561 0710**

reception@cleanpower.com.au
www.cleanpower.com.au

David Brockway **Chief Executive Officer**
Pete Saunders **Chairman**

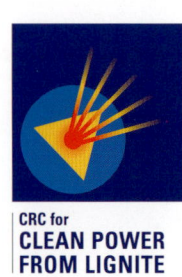

CRC DIRECTORY

ABN. 62 089 499 034   See page 86 for our Case Study

COOPERATIVE RESEARCH CENTRE FOR THE **GREAT BARRIER REEF WORLD HERITAGE AREA**

Located at Sir George Fisher Building, James Cook University, Townsville QLD 4811
Phone. **07 4781 4976**   Fax. **07 4781 4099**

crcreef@jcu.edu.au
www.reef.crc.org.au

Dr Russell Reichelt **Chief Executive Officer**
Sir Sydney Schubert **Chairman of the Board**

ABN. 52 060 244 537   See page 121 for our Case Study

COOPERATIVE RESEARCH CENTRE FOR **INTELLIGENT MANUFACTURING SYSTEMS**

Cnr Albert & Raglan Streets, Preston VIC 3072
PO Box 348, Preston VIC 3072
Phone. **03 9480 0400**   Fax. **03 9480 0413**

info@crcimst.com.au
www.crcimst.com.au

Mr David Galvin **Executive Director**
Professor Brian Smith **Chairman of the Board**

ABN. 81 470 130 610   See page 102 for our Case Study

COOPERATIVE RESEARCH CENTRE FOR **MOLECULAR PLANT BREEDING**

Waite Building, Waite Campus, The University of Adelaide, PMB 1 Glen Osmond, South Australia 5064
Phone. **08 8303 6539**   Fax. **08 8303 6789**

crcmpb@waite.adelaide.edu.au
www.molecularplantbreeding.com

Bryan Whan **Director**
Ian Atkinson **Business Manager**

Cooperative Research Centre Molecular Plant Breeding

ABN. 71 085 103 528   See page 124 for our Case Study

REDFERN PHOTONICS PTY LTD

Suite 212 National Innovation Centre, Australian Technology Park, EVELEIGH NSW 1430, Australia
Phone. **02 9209 4900**   Fax. **02 9209 4910**
enquiry@redfernphotonics.com.au   www.redfernphotonics.com

AUSTRALIAN PHOTONICS CO-OPERATIVE RESEARCH CENTRE

Suite 201 National Innovation Centre, Australian Technology Park, EVELEIGH NSW 1430, Australia
Phone. **02 9351 1901**   Fax. **02 9351 1910**
info@photonics.com.au   www.photonics.com.au

AUSTRALIAN SCIENCE

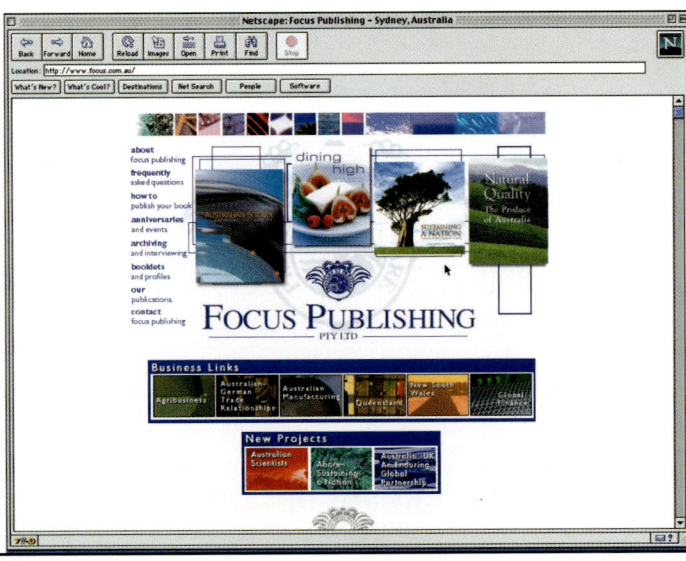

www.focus.com.au

Visit the Focus website for information on Focus Publishing, Australia's leading corporate book publisher. Focus specialises in producing high-quality custom books, corporate histories and specific marketing, event, promotional and anniversary books.

To obtain further information on the companies and organisations participating in this book, *Australian Science—Expanding the Frontiers*, simply follow the steps.

> Enter the Focus Website address: www.focus.com.au
> You are now on Focus' home page. Click on the words our publications.
> Scroll down to the bottom of the page and click on latest publications.
> Scroll down the page to find *Australian Science—Expanding the Frontiers.*
> Click on Business Links. You will find the websites of the participating companies and other organisations listed in light blue type.
> Click on your chosen company's or organisation's address and it will lead you to their website.

WEBSITE

Anti-Cancer Council of Victoria
www.accv.org.au

Allens Arthur Robinson
www.arh.com.au

AstraZeneca Pty Ltd
www.AstraZeneca.com.au

Australian Photonics Co-operative Research Centre
www.photonics.com.au

Austin Research Institute
www.ari.unimelb.edu.au

Australian Pharmaceutical Manufacturers Association
www.apma.com.au

Australian Proteome Analysis Facility
www.proteome.org.au

Australian Research Council
www.arc.gov.au

Australian Technology Park
www.atp.com.au

Baker Medical Research Institute
www.baker.edu.au

Biota Holdings Limited
www.biota.com.au

Centenary Institute of Cancer Medicine and Cell Biology
www.centenary.usyd.edu.au

Children's Cancer Institute Australia for Medical Research
www.ccia.org.au

CRC for Advanced Composite Structures Ltd
www.crc-acs.com.au

CRC for the Biological Control of Pest Animals
www.pestanimal.crc.org.au

CRC for Black Coal Utilisation
(Centre for Coal in Sustainable Development)
www.newcastle.edu.au/department/black_coal_crc

CRC Reef Research Centre
www.reef.crc.org.au

CRC for Intelligent Manufacturing Systems and Technologies Ltd
www.crcimst.com.au

CRC for Molecular Plant Breeding
www.molecularplantbreeding.com

CRC for Clean Power from Lignite
www.cleanpower.com.au

CSL Limited
www.csl.com.au

Eli Lilly Australia Pty Limited
www.lilly.com.au

AUSTRALIAN SCIENCE

Garvan Institute of Medical Research
www.garvan.unsw.edu.au

GlaxoSmithKline Australia
www.gsk.com.au

Hanson Centre for Cancer Research
www.imvs.sa.gov.au/hanson

Howard Florey Institute
www.hfi.unimelb.edu.au

John Curtin School of Medical Research
www.jcsmr.anu.edu.au

Macfarlane Burnet Centre for Medical Research
www.burnet.edu.au

Mater Medical Research Institute
www.mmri.mater.org.au

Menzies School of Health Research
www.menzies.edu.au

Merck Sharp & Dohme
www.msda.com.au

Michael Johnson & Associates Pty Limited
www.mjassoc.com.au

Monash University
www.monash.edu.au

Murdoch Childrens Research Institute
www.murdoch.rch.unimelb.edu.au

Optiscan Pty Ltd
www.optiscan.com

PricewaterhouseCoopers
www.pwcglobal.com/autice

Prince Henry's Institute of Medical Research
www.med.monash.edu.au/phimr

Prince of Wales Medical Research Institute
www.powmri.unsw.edu.au

Redfern Photonics Pty Ltd
www.redfernphotonics.com

State Government of Victoria—Science, Technology & Innovation
www.innovation.vic.gov.au

St. Vincent's Institute of Medical Research
www.svimr.unimelb.edu.au

Starpharma Pooled Development Limited
www.starpharma.com

The Mental Health Research Institute of Victoria
www.mhri.edu.au

The Queensland Institute of Medical Research
www.qimr.edu.au

The University of Melbourne
www.unimelb.edu.au

The University of Sydney
www.usyd.edu.au

The Walter and Eliza Hall Institute of Medical Research
www.wehi.edu.au

Thomson Marconi Sonar
www.tms-pty.com

TVW Telethon Institute for Child Health Research
www.ichr.uwa.edu.au

INDEX

AAMRI 45, 169
Aborigines 61
Academy of Science, Australian 18
actuators 118
Ada, Gordon 22
Adacel 126
Advanced Micro Devices 78
agribusiness 149
agriculture 9, 20, 28, 81, 99-105
AJ Parker CRC for Hydrometallurgy 113
algal blooms 88
Allens Arthur Robinson 83, 170
Alzheimer's Disease 44, 59
AMRAD Corporation 78, 88, 162
angiogenesis 79
Anglo-Australian Planet Search 135-36
Anglo-Australian Telescope (AAT) 32-33, 133-34, 136-37
animal diseases 103-4
Animal Gene Storage Research Centre of Australia (AGSRCA) 82
Antarctica 19-20, 138
Anti-Cancer Council of Victoria 63, 169
antivenenes 76
anxiety 44-47
Apollo missions 33
apoptosis 58
Application Development Automatic Generation Environment (ADAGE) 126
Applied Science Associates (ASA) 88
aquaculture 86-87
Arid Zone Recovery Project 95
asthma 50, 162
AstraZeneca 56, 64, 154-57, 170
astronomy 15, 27-33, 133-38, 151
astrophysics 138
atherosclerosis 43, 51
ATM technology 130
atom bomb 26, 27
atomic absorption spectrophotometer 28
Ausmelt 107-8
Auspace 137
Austin Research Institute (ARI) 42, 171
Australasian Association for the Advancement of Science 17
Australia Prize 8
Australia Technology Showcase of New South Wales 9
Australia Telescope (AT) 134, 137
Australian Academy of Science 18
Australian Animal Health Laboratory (AAHI) 103-4
Australian Atomic Energy Commission 26
Australian Coal Association Research Program 110
Australian Defence Scientific Service 27
Australian Diabetes Survey 162
Australian Genome Research Facility (AGRF) 76
Australian Institute for Marine Science (AIMS) 85, 87
Australian Membrane and Biotechnology Research Institute (AMBRI) 82
Australian Mineral Industries Research Association (AMIRA) 114
Australian National University (ANU) 24, 27, 32-33, 39, 42, 65, 79, 135, 137
Australian Pharmaceutical Manufacturers Association (APMA) 62, 171
Australian Proteome Analysis Facility (APAF) 76, 77, 171
Australian Research Council (ARC) 9, 172
Australian Resources Research Centre 108, 113
Australian Technology Park Innovations (ATPI) 37-38, 39, 172
automation research 111-12
aviation 29
AWA 126

Backing Australia's Ability plan 36
bacteriology 22
Baker Medical Research Institute 51, 172
Bakoss, Professor Steve 38
Bancroft, Thomas 24
Banks, Joseph 15
Bauer, Ferdinand 16
Beard, Maston 26-28
Berkovic, Professor Sam 43
bilby 95
BioDiscovery 75
biodiversity 56, 96
biological pest control 102
Biomechanics Research Laboratory 118
biomedical research 58
biomonitoring fabrics 117-18
Bionic Ear Institute 125
biopharmaceuticals 78
biosensor 80-82
Biota 83, 173
biotechnology 75-83, 87-88, 148
Biotron 43
blow-fill-seal technology 64
Bolton, John 31
Boot, Henry 27
botany 15
Bowen, EG ('Taffy') 31, 32
BP Solar 93
bra 118
brain 43
Brand-Miller, Professor Jennie 54-57
breast cancer 53
BresaGen 57, 78
Brisbane, Sir Thomas 9
broadband networked telecommunications 130
Broken Hill (NSW) 107
Broken Hill Proprietary Ltd 18
Brown, Robert 9, 16
Bureau of Mineral Resources 21, 26
Burnet, Sir Frank Macfarlane 22, 38
business services 37, 83, 112

Cadeceus computer 78
calicivirus 94
cancer
   biotechnology 79
   breast 53
   cervical 67
   childhood 52
   diagnosis 70, 76
   drugs for 47
   leukemia 57
   melanoma 71, 78
   research into 50, 63
   role of inhibin 49
   vaccines 42, 60, 66, 67, 71
cattle breeding 103
CDMA 130
Centenary Institute of Cancer Medicine and Cell Biology 50, 173
Centre for Advanced Technology in Telecommunications (CATT) 126
Centre for Health Informatics (CHI) 123-24
Centre for Photovoltaic Engineering 92-93
Centre for Quantum Computing Technology 131
Centre for Telecommunications and Industrial Physics (CTIP) 128
cervical cancer 67
Cerylid 75
Chain, EB 23
chemistry 28
Chief Scientist, first 8
childhood disease 52, 54
Children's Cancer Institute Australia for Medical Research 52, 173
circumnavigation 16

citrus industry 101
Clark, Professor Bob 131
Clarke, Rev. William Branwhite 21
climate forecasting 105
clinical communications 124
Clinical Outcomes and Research Institute 43
clinical research trials 43
cloning 48, 49
clothing, health-monitoring 118
Clunies Ross Awards 9
Clunies Ross, Ian 28
coal 109, 111
Comforth, John 28
Commonwealth Scientific and Industrial Research Organization see CSIR/CSIRO
Comprehensive Cancer Research Centre 71
computer chips 92
computers
   development 28
   ocean current modelling 89
   pharmaceutical design 78
   research into 131
Condor 126
construction industry 120
continental drift 21
Cook, James 15-16, 133
Cooke, Dr Brian 94
Cooperative Research Centres 8, 36-37, 140-43
   Advanced Composite Structure (CRC-ACS) 118, 184
   Asthma 162
   Biological Control of Pest Animals 95, 184
   Black Coal Utilisation 109, 174, 184
   Cellular Growth Factors 77, 78
   Clean Power from Lignite 115, 174, 185
   Discovery of Genes for Common Diseases 75
   Dryland Salinity 94
   Eye Research and Technology 76
   Great Barrier Reef World Heritage Area 85, 185
   Hydrometallurgy 113
   Intelligent Manufacturing Systems and Technologies Ltd 121, 185
   Mining Technology and Equipment 110-11
   Molecular Engineering and Technology 82
   Molecular Plant Breeding 102, 185
   Photonics 124, 184
   Satellite Systems 137
   Telecommunications 130
   Vaccine Technology 163
coral 87
Cornell, Dr Bruce 8, 82
Cory, Suzanne 23
cosmic rays 136
cosmic studies 30
cotton industry 100
Cox, AB 28
CRCs see Cooperative Research Centres
CSIR/CSIRO 26-27
   Building, Construction and Engineering Division (BCE) 119-20
   Division of Radiophysics 18
   Energy Technology 110
   Entomology 75
   Human Nutrition 54
   Land and Water 94
   Livestock Industries 103
   Marine Science 85
   Microalgae Research Centre 88
   mineral science 108

Mining 113
Molecular Science 119
myxoma trials 25
Petroleum and Exploration 113
Plant Industry 99
radio astronomy 30-31
telecommunications 127-28, 129-30
Telecommunications and Industrial Physics 92
CSIRAC computer 28
CSL 66, 163, 174
currents, ocean 89
Curtin University of Technology 113, 130
cutting technology 110-11
cytokines 58

dairy industry 104
Dampier, William 16
Darwinism 17
David, TW Edgworth 19
Dean, Professor Brian 44
Defence Services Technology Organisation (DSTO) 126
defence technologies 164-65
dendrimers 69
dendritic cells 60
dengue fever 24
Dennis, Dr Liz 8, 99
designer tissue 48-49
diabetes 47-48, 50
diagnosis
   biosensor 82
   cancer 70, 76
digital certificates 125-26
DNA sequencing 76
Doherty, Professor Peter 42
Donoghue, Dr Neil 80
Douglas Mawson Telescope (DMT) 138
Droege, Professor Peter 93
Dunlap, Dr Walt 88

Earth Sanctuaries Limited 95-96
Easteal, Dr Simon 53
e-business 125-26
Eccles, John 25
ecology 91-97
Edman, Pehr 47
education 17-18
Elcom Technology 39
Eli Lilly Australia 43, 175
Ellery, Robert 29
embryonic stem cells 68
endangered animals 82, 94, 95-96
energy 10, 92-94, 119, 142
environment 91-97, 143
enzymes 28
epilepsy 43, 53
Epilepsy Research Institute 43
Ericsson Australia 126, 130
ethical issues 146-47
Eureka (prize) 8
European Southern Observatory (ESO) 138
Eurosolare 93
evolution 150
exploration
   Antarctic 19-20
   Australia 15-16, 17

fabrics, biomonitoring 117-18
Farrer, William 20
Federation Satellite One (FedSat-1) 137
Fenner, Frank 22
fertility 49
fertility control agents (pest management) 95
fission physics 27
flight engineering 29

189

AUSTRALIAN SCIENCE

Flinders, Matthew 16
Florey, Howard 23-24
Flowering Switch Gene 99-100
Floyd, Dr John 108
food technology 81
Frazer, Professor Ian 67
French scientists 16
fusion physics 27
future 148

Gadjusek, Carleton 22
Gaensler, Dr Brian 9
Gage, Professor Peter 42
Garvan Institute of Medical Research 44, 162, 175
Gemini project 136-37
gene chip technology 50, 53, 65
GeneSTAR Marbling 103
genetic engineering 81
Genetic Solutions 103
genetically modified foods 81
genetics; see also biotechnology
    agriculture 99-103, 102-3
    ethics 147
    health problems 65
    Mendel 20
    therapy 38
geological surveys 17, 21
geology 20-21
'GI' symbol 57
GigaWave 124
Giles, Professor Graham 63
glass 32
Glass Earth concept 108
GlaxoSmithKline 55, 80, 175
Global Positioning System (GPS) 129-30
glycaemic index 57
GM foods 81
gold mining 21, 114
grains industry 102
grants 9
Great Barrier Reef 10, 56, 85, 87
Green, Professor Martin 10, 92
Griffith University 56
GSAO drug 80

Hanson Centre for Cancer Research 57, 176
Hargrave, Lawrence 29
Hawke, Bob 8
health care 124-25, 150; see also medical research
heart disease 43, 47, 51
helium 27
Hicks, Professor Rodney 50
Hill, Professor David 63
HIV 42, 77, 79-80
Hogg, Professor Philip 80
Holloway, Dr Andrew 50
Hollows, Fred 25
home telecare project 124-25
horticulture 101
Horticulture Research and Development Corporation 101
Hospital Without Walls 57
Howard Florey Institute of Experimental Physiology and Medicine 24, 53, 176
human genome 58, 147
Human Genome Project 58
human papilloma virus (HPV) 67, 162
hypertension 51, 162

immunology 21, 22, 24-25, 66, 71, 149, 163
in vitro fertilisation (IVF) 24
Indian Pacific Communications 39
Industrial, Technological and Sanitary Museum (Sydney) 17
influenza drug 78, 83

information technology 143
inhibin 49
injury prevention 54
Innovation Summit Implementation Group 36
Innovations Statement 7
insect control 100
Institute for Molecular Bioscience (IMB) 79
Institute of Drug Technology 80
Institute of Reproduction and Development 68
Institute of Science and Industry 28
integrated pest management (IPM) 101
intellectual property 147
Intelligent Polymer Research Institute (IPRI) 117-19
Interactive Voice Response (IVR) 125
International Energy Agency 93
Internet 126, 127, 130
ion channels 42-43
Iscom 66
IVF (in vitro fertilisation) 24

James Cook University 85, 87
John Curtin School of Medical Research (JCSMR) 23-24, 42, 53, 65, 176
Joint Australian Centre for Astrophysical Research on Antarctica (JACARA) 138
Jones, Barry 8
Julius Kruttschnitt Mineral Research Centre 114

Kellaway, Charles 22
Kenny, Elizabeth 24-25
Kimberley region 10, 48
Kirby, Justice Michael 146-47
Knowledge Nation 7
Kuhse, Helga 146-47
kuru 22

land repair 94-95
laser plasma spectrometer (LPS) 115
law services 83
leukemia 57
light harvesting polymers 119
Liversidge, Archibald 17
Lopez, Professor Angel 57
lunar radiation 30

McCoy, Frederick 17
Macfarlane Burnet Centre for Medical Research 46, 177
McGee, Dick 31
Mackay, Alistair 19
Macnamara, Jean 24
Macquarie University 76, 77, 130
Madsen, John 30
magnesium 114
Major National Research Facility scheme 76
malaria 75
Manhattan project 27
manufacturing CRCs 141
Maralinga 26
Marine Bioproducts 88
marine science 85-89
Marshall, Barry 77
Martin, Professor Donald 43
Martin, Professor Nicholas 44-47
Martyn, David 18, 30
Masters, Professor Colin 44
Mater Hospital Research Institute 60, 177
materials 10, 115, 117-21
Mawson, Douglas 19-20
Medical Genome Centre 65
medical research 10, 21-25, 41-72, 142; see also health care
melanoma 70, 78
Memcor process 92

mental health 44, 45, 53, 59
Mental Health Research Institute (MHRI) 44, 59, 181
mental retardation research 44
Menzies School of Health Research 61, 177
Merck Sharpe & Dohme 67, 162, 178
metallurgy 28
metals technology 114
Metcalf, Donald 58
Michael Johnson & Association (MJ&A) 37, 178
Miles, David 36
Mills, Dr David 93
mineral resources 10, 20-21, 107-15
minerals processing 113, 114
mining 26, 27, 107-15, 142
Minnett, Harry C 30
mobile phones 124-25
modems 130
Molonglo Observatory Synthesis Telescope (MOST) 135
Monash University 49, 68, 178
moon landing 33
Moore, Neil 24
Motorola Australian Research Centre (MARC) 127
Mt Stromlo observatory 32
mouse library 53
Mueller, Ferdinand Jakob Heinrich von 18, 19-20
munitions research 32
Murdoch Children's Research Institute 82, 179
Murray-Darling Basin 94
museums 17
myxoma virus 25, 28, 94

nanotechnology 80-82, 92, 120
Narrabri Stellar Intensity Interferometer 134
NASA 33
nasal continuous positive airway pressure device (CPAP) 57
National Trachoma and Eye Health Program 25
natural history 10, 16-17
Near-infrared Integral-Field Spectrograph (NIFS) 137
networks, telecommunications 130
NSW Retinal Dystrophy Research Centre 72
nickel 113
Nobel Prize
    for Chemistry 28
    for Medicine 22, 23, 25, 38
Nossal, Gustav 22-23
nuclear science 26-27
nutrition 54, 54-57, 149

observatories 29, 32-33
ocean currents 89
oceanography 85-89
oil and gas industry 89
oil spill management 89
Oilmap 89
Oliphant, Marcus 18, 27
ophthalmology 25, 72
Optical Munitions panel 32
Optiscan Imaging 70, 179
organic chemistry 28

P9 Project 114
Pacific Power 94
Pacific Solar 93
pain control 78
parasitology 24
Parish, Professor Chris 79
Parker Centre II 113
Parkes radio telescope 32, 134

Parkinson, Sydney 15
Parkinson's Disease 48
Paul Wild Observatory 134
Pawsey, Joseph Lade 32
Peacock, Dr Jim 8, 99
Pearcey, Trevor 26-28
penicillin 23, 162
Pera, Dr Martin 49, 50
pest control 93, 95, 100, 101
Peter MacCallum Cancer Institute (PMCI) 50, 53
pharmaceutical product packaging 64
pharmaceuticals 62, 78-80, 88, 162-63
photosynthesis, artificial 92
photovoltaic technology 92, 119, 151
physics 27
PI 88 drug 79
Piddington, Jack 30
pipeline materials 120
plant breeding 20
plant genetics 99-100, 149
plutonium 26-27
policy, science 7-8
poliomyelitis 24-25
polymers 119
Pondman 87
positron emission tomography (PET) 50
potato industry 101
prawn farming 87
PriceWaterhouseCoopers (PwC) 112, 179
Prime Minister's Prize for Science 8, 99
Prince Henry's Institute of Medical Research 49, 180
Prince of Wales Medical Research Institute (POWMRI) 43, 48, 54, 180
prizes 8-9; see also Nobel Prize
protein X-ray crystallography 47
proteomics 76-77
Pure Commerce 39

quantum computing 131
qubit devices 131
Queensland Centre for Advanced Technologies (QCAT) 108-9, 108-10
Queensland Centre for Climate Applications 105
Queensland Institute of Medical Research (QIMR) 44-47, 71, 182
Queensland Institute of Molecular Biosciences 103

R&D support consultancy 37
rabbit control 25, 94
radar 18, 27, 30
radiation 18, 30-31
radio astronomy 30-32, 134, 137-38
Radio Frequency Systems 130
radioactivity 26-27
Radiophysics Laboratory 30-31
radium 20
rainforests 56
Ramshaw, Professor Ian 79
Randall, John 27
reagents 28
Redfern Photonics 124, 184
Relenza 78, 83
reproductive technologies 82, 147
Research School of Astronomy and Astrophysics (RSAA) 137
ResMed 57
resonant cavity magnetron 27
resources, mineral 20-21, 107-15
Rivett, David 30
RMIT University 126
Royal Adelaide Hospital 57
Royal Botanic Gardens (Melbourne) 18
Royal Societies 18
Rum Jungle 26

INDEX

St. Vincent's Institute of Medical Research 47, 181
salinity 94-95
Salk vaccine 24
SAMAG 114
satellite 137
schizophrenia 44, 53
Science Council 8
science policy 7-8
sensors 118
serial analysis of gene expression (SAGE) 53
sewage treatment 92
sexually transmitted diseases 69
Siding Spring observatory 32-33
Simpson, Dr Ann 47
Sirosmelt Top Submerged Lance (TSL) 107-8
skin cancer 70, 78
Slatyer, Professor Ralph 8
sleeping sickness 80
smart textiles 117-18
Smartex 118
smelting 107-8
Smith, Dan 101
snoring prevention 57
societies, scientific 17-18
Solander, Daniel Carl 15
Solar City program 93
solar energy 92-93
solar radiation 18, 30
sonar 164-65
south pole 19
space industry 9-10, 137
spectrometer 115
speech technology 127
Spencer, Walter Baldwin 17
spina bifida 54
spinal injury 54

Spring, Herman 15
Square Kilometre Array (SKA) 138
Stanley, Professor Fiona 54
Starlab project 137
Starpharma Pooled Development 69, 181
stem cells 68
Stone, Professor Jonathon 72
sugar industry 101-2
suicides 59
Sullivan, Professor Colin 57
sunscreen, marine 88
Sunscreen Technologies P/L 88
Suppressors of Cytokine Signalling (SOCS) proteins 77
SureTRAK 103
surface science 119
Sutherland, Professor Grant 43-44
Sutherland, Professor Struan 76
Switched Network Research Centre 126
Sydney University Stellar Interferometer (SUSI) 134
Symon, Professor Bob 78
synapses 25

Tebbutt, John 9, 27-29, 133
TECRA 118
telecommunications 10, 123-31, 143
Telecommunications Information and Technology Centre (TITC) 126
telehealth 57
telephone services 125-26
telescopes 31-33, 133-35, 138
Telstra 125-26
thermonuclear weapons 27
Thomas Marconi Sonar 164-65, 183
tissue, designer 48-49
toilet cisterns 92
tourism 85

trachoma 25
transgenesis 65
transplantation 42
Trounson, Professor Alan 49, 50, 68
trypanosomiasis 80
tumours see cancer
TurboFlotation 110
TVW Telethon Institute for Child Health Research 54, 183

university education 17
University of Adelaide 44, 136
University of Melbourne 17, 24, 28, 38, 44, 76, 182
University of New South Wales (UNSW) 39, 76, 80, 92-93, 123
University of Queensland 76, 78, 114
University of Sydney 17, 39, 54-57, 72, 93, 97, 134, 158-61, 182
University of Technology, Sydney (UTS) 39, 43, 47
University of Western Australia 54
University of Wollongong 117, 126
uranium 26

vaccines
  for cancer 42, 60, 66, 67, 71, 163
  for HIV 79-80
  for HPV 67, 162
Very Long Baseline Interferometry (VLBI) 134
veterinary products 78
Victorian Government 79, 180
VIR 201 vaccine 79
Virax Holdings 79
virology 22, 42
visual image processing 127
Vivendi Water 92

Vodafone Network 130
von Mueller, Ferdinand Jakob Heinrich 18, 19

Walsh, Alan 28
Walter and Eliza Hall Institute of Medical Research (WEHI) 21, 38, 58, 76, 183
Wamsley, Dr John 95-96
WAP (wireless application protocol) 124
Wark, Ian 28
Warren, Robin 77
water management 92, 94-95, 100-101
weapons, thermonuclear 27
weather science 105
Western Australian Biomedical Research Institute (WABRI) 80
WA Chemistry Centre 80
wheat 20, 102
Williams, Dr Keith 76
Williamstown Observatory 29
wind power projects 94
wireless application protocol (WAP) 124
wireless program research 130
Women and Children's Hospital (Adelaide) 44
Wood, Ian 22
Woolley, Richard van der Riet 32-33
Woomera 27
World War I 28
World War II
  medical research 23
  nuclear science 26
  radio astronomy 30, 32
World Wide Web 124-25

zinc 28
Zinkernagel, Rolf 42

>> ABOUT THE WRITERS

LEIGH DAYTON is a science writer and broadcaster specialising in science, technology, the environment and medicine, and has recently been awarded The 2001 Industry, Science and Technology Michael Daley Eureka Prize for Science Journalism. She studied at the University of California and the University of Alberta in Canada. In Canada, she produced David Suzuki's *Discovery* for national radio, plus three other award-winning radio series. She worked with the National Research Council and Simon Fraser University, and wrote for the London *Daily Telegraph*, Canada's *Globe and Mail* and *New Scientist* magazine. In 1990, she moved to Australia, writing for *New Scientist*, *Omni* magazine and the *Sydney Morning Herald*, where she won four journalism prizes. She has also appeared on the *Today Show* (Channel 9) and ABC television and radio.

JOHN O'BYRNE (BSc PhD, Sydney) is a Senior Lecturer in Physics at the University of Sydney. He has worked in Australia and the United States exploring his interests in achieving the best possible resolution using ground-based optical telescopes. He is a Secretary of the Astronomical Society of Australia, the society of professional astronomers in Australia. As well as lecturing in physics to undergraduate students, he has for many years conducted astronomy courses for adults and, in recent times, he has been an author or consultant editor for several popular astronomy books for readers of all ages.

GORDON COLLIE is an award-winning Brisbane writer and consultant. A former New Zealander, he spent 11 years reporting with Queensland *Country Life* before joining the Brisbane *Courier-Mail* in 1985. He is the only person to win the Dalgety Award for Excellence in Rural Journalism in Queensland three times. He also travelled and wrote extensively from South-East Asia while on an ASEAN Journalism Scholarship. In 1999 he was a founding partner in the specialist rural communications company Agri-Prose, which has clients in the commercial, government and rural sectors. He assumed the role of managing director in May 2000.

AMARA BAINS is a health/science writer and communications consultant, currently writing for trade publications in health and science and for the pharmaceutical and medical industry. She has spent ten years working in the healthcare area. The first five of those years were spent in the pharmaceutical industry, the latter five in marketing and public relations. Amara has a science degree from the University of Sydney, and has completed postgraduate studies in marketing and public relations and a Masters of International Public Health. While she specialises in healthcare, Amara is also involved in the arts, and has worked for Bangarra Dance Theatre.

BERNADETTE HINCE is an Australian who loves words, and is at a loss without a dictionary. She has worked on the *Australian National Dictionary* and the *Dictionary of New Zealand English*, and is the author of *The antarctic dictionary* (CSIRO/Museum Victoria 2000), an 11-year project which chronicles the vocabulary of the south. Bernadette has academic qualifications in English and science, and was the 1995 Thomas Ramsay Science and Humanities Fellow at the Museum of Victoria. Her interests include Australian natural history and science, people's use of plants, historical botany and the polar regions. She is currently writing an environmental history of the subantarctic islands.

A Focus Publishing Book Project
Focus Publishing Pty Ltd
ABN 55 003 600 360
PO Box 518 Edgecliff NSW 2027
Telephone 61 2 9327 4777
Fax 61 2 9362 3753
Email focus@focus.com.au
Website www.focus.com.au

PROJECT TEAM
Project Manager. Andra Müller
Commissioning Editor. Susan O'Flahertie
Sub Editor. Sarah Shrubb, Kevin Pyle
Designer. Sarah Cory
Client Services. Kate Sanday, Karen Tyrrell
Corporate Communications. Gloria Nykl
Production Manager. Timothy Ho

Chairman. Steven Rich AM
Publisher and CEO. Jaqui Lane
Associate Publisher. Gillian Fitzgerald
Managing Editor. Philippa Sandall

© 2001 Focus Publishing
This book is copyright. Apart from any fair dealing for the purpose of private study, research, criticism or review, as permitted under the Copyright Act, no part may be reproduced by any process without written permission.
Enquiries should be addressed to the publisher.
Whilst all reasonable attempts at factual accuracy have been made, Focus Publishing accepts no responsibility for any errors contained in this book.

ISBN 1 875359 78 8

For information on Focus Publishing visit www.focus.com.au